Methods of Optimization

Methods of Optimization

G.R. WALSH

Department of Mathematics,
University of York

A Wiley–Interscience Publication

JOHN WILEY & SONS
London · New York · Sydney · Toronto

Library of Congress Cataloging in Publication Data;

Walsh, Gordon Raymond, 1925–
Methods of optimization.

'A Wiley-Interscience publication.'
Bibliography: p.
Includes index.
1. Mathematical optimization. I. Title.
QA402.5.W34 515′.84 74-20714

ISBN 0 471 91922 5 (Cloth)

ISBN 0 471 91924 1 (Pbk)

Photosetting by Thomson Press (India) Limited, New Delhi, and printed by The Pitman Press, Bath, Avon, England.

Preface

This book is concerned with modern methods of maximizing or minimizing a given function of many real variables. It will be seen from the list of references that most of these methods have been developed over the last two decades, 1950–1970. Since new methods and new combinations of old methods are appearing almost daily, the choice of material is largely a matter of personal preference. It is hoped, however, that all the fundamental ideas in modern optimization theory have been included and that every important type of method is represented. The major aim of this book is to present these methods in a straightforward and systematic way. Wherever possible, an historical approach has been used, for it is interesting to see why, as well as how, certain methods were developed.

Apart from the intrinsic interest of the subject, the study of optimization techniques is attractive because of its very wide field of applications. Optimization problems arise in operational research, economics, aerospace, pure geometry, physics, control theory, chemical engineering and in many other subjects. It has even been claimed that *everyone* is optimizing *something* all the time!

The methods described here are confined to what is usually called 'static optimization' or 'parameter optimization'. This means that the functions to be optimized depend on variables which all have the same status—there are no special variables such as 'time', 'control', etc. The introduction of such variables would lead to a study of optimal control theory which, to do it justice, requires a separate treatment. Similarly, any constraints on the variables are algebraic equations or inequalities, rather than differential equations.

The text is based on a course of lectures given to third-year mathematics undergraduates in the University of York. The exercises at the ends of the chapters are intended to illustrate the theory in a simple way—an exercise involving two variables is often just as useful as a similar exercise involving ten variables (and is sometimes more useful). Most of the numerical exercises can be solved without the aid of any computing machinery, and almost all the others require only a small desk machine. Students are advised to write their own computer programs for the various algorithms, for, apart from the satisfaction of seeing that one's own versions of the algorithms really work, such programs are relatively easy to amend when necessary.

Some standard linear programming theory and practice is needed from time to time in Chapters 1 and 2. It is assumed that the reader is familiar with the elements of this subject, as set out, for example, in the author's *An Introduction to Linear Programming.*[75] Otherwise, the only prerequisites are a basic knowledge of the differential calculus of several variables and of the elementary theory of vectors and matrices.

Chapter 1 begins with a classification of the various optimization problems considered in the book. Sections 1.2 to 1.4 deal with classical optimization, its extensions and limitations, and the significance of Lagrange multipliers. The chapter ends by defining and discussing convex and concave functions.

Chapter 2 is largely concerned with the Kuhn–Tucker theory of nonlinear programming. This theory is developed in detail and is then applied to the quadratic programming problem. In the final section of this chapter, an interesting and useful alternative method for solving nonlinear programming problems is derived, namely, the method of Griffith and Stewart.

Chapter 3 provides a selection of search methods for unconstrained optimization, most of them direct search methods. For later use, the linear search methods of Powell (quadratic interpolation) and Davidon (cubic interpolation) are included.

Chapter 4 gives a systematic development of the various gradient methods for unconstrained optimization, starting from the simplest—the steepest descent method—and proceeding to the more elaborate conjugate direction methods, including those for which the evaluation of derivatives is not necessary.

Chapter 5 shows how constraints can be included in the algorithms of the previous two chapters. It is fortunate for optimization theory that constraints can be dealt with in this isolated way.

Chapter 6, on dynamic programming, is entirely independent of the rest of the book. It gives a brief introduction to the subject, together with some typical examples of the dynamic programming technique.

It is again a pleasure to thank my colleague, Dr. N. Anderson, for his careful reading of an early draft of this book and for his many excellent comments. I would also like to thank my 'optimization' students for constructing some of the exercises, for verifying the answers to many others and for pointing out several obscurities.

G.R. WALSH

Notation

The point in Euclidean n-space E^n with coordinates $[x_1, \ldots, x_n]$ is denoted by the column vector $\mathbf{x}$ and a function $f(x_1, \ldots, x_n)$ is denoted by $f(\mathbf{x})$.

Vectors and matrices are printed in bold type, with a dash denoting the transpose of a vector or matrix. Vector inequalities should be interpreted component by component.

Unless otherwise stated, the suffix i takes the values $1, \ldots, m$ and the suffix j takes the values $1, \ldots, n$. For example, $g_i(\mathbf{x}) = b_i$ means

$$g_1(\mathbf{x}) = b_1, \ldots, g_m(\mathbf{x}) = b_m.$$

The summation symbols $\sum_i$ and $\sum_j$ will always mean $\sum_{i=1}^{m}$ and $\sum_{j=1}^{n}$ respectively.

Contents

CHAPTER 1

Introduction

1.1. WHAT IS OPTIMIZATION?

Tout est pour le mieux dans le meilleur des mondes possibles. Voltaire

The optimist proclaims that we live in the best of all possible worlds; and the pessimist fears this is true. J. B. Cabell

The fundamental problem of optimization is to arrive at the best possible decision in any given set of circumstances. Of course, many situations arise where the 'best' is unattainable for one reason or another; sometimes what is 'best' for one person is 'worst' for another; more often we are not at all sure what is meant by 'best'. The first step, therefore, in a mathematical optimization problem is to choose some quantity, typically a function of several variables, to be maximized or minimized, subject possibly to one or more constraints. The commonest types of constraint are equalities and inequalities which must be satisfied by the variables of the problem, but many other types of constraint are possible; for example, a solution in integers may be required. The next step is to choose a mathematical method to solve the optimization problem; such methods are usually called optimization techniques, or algorithms.

The choice of optimization technique is by no means obvious, for the theory and practice of optimization has developed rapidly since the advent of electronic computers in 1945. It came of age as a subject in the mathematical curriculum in the 1950's, when the well-established methods of the differential calculus and the calculus of variations were combined with the highly successful new techniques of mathematical programming which were being developed at that time. The programmers, it was said, had joined forces with the hillclimbers.

The optimization problems that have been posed and solved in recent years have tended to become more and more elaborate, not to say abstract. Perhaps the most outstanding example of the rapid development of optimization techniques occurred with the introduction of dynamic programming by Bellman in 1957 and of the maximum principle by Pontryagin in 1958. These techniques were designed to solve the problem of the optimal control of dynamical systems. Both dynamic programming and the maximum principle are closely related to the calculus of variations, and hence to each other. A brief introduction to discrete dynamic programming is given in Chapter 6; the maximum principle and optimal control theory are beyond the scope of this book.

The simply-stated problem of maximizing or minimizing a given function of several variables has attracted the attention of many mathematicians over the past twenty-five years or so. The direct search methods of solution, which involve function evaluations and comparisons only, are usually simpler, though less accurate for the same computational effort, than the indirect or gradient methods, which require values of the function and its derivatives. Both types of method are still undergoing development, with the major emphasis being on the search for efficient and reliable algorithms to deal with general nonlinear functions.

For the sake of simplicity, and following the historical development of the subject, most of the theory of the gradient methods in this book is restricted to the case of quadratic functions. The theory for non-quadratic functions is considerably more difficult and is the subject of current research; it is therefore unsuitable for a first course in optimization. However, the algorithms based on quadratic theory are usually successful when applied to non-quadratic functions.

The following examples are typical of optimization problems that arise in practice.

Statistics. The frequency function of a population is completely determined once its parameters are known. For example, the binomial distribution is completely determined by the parameters n (the number of independent trials of an experiment) and p (the probability of success in a single trial). An important problem in statistics is to estimate the population parameters, given a random sample drawn from the population. If the form of the frequency function is assumed, then values for its parameters may be determined by forming the *likelihood function*, which gives the probability that the given sample came from a population with the assumed frequency function. The likelihood function is thus a function of the unknown parameters. The values of the parameters are now estimated by maximizing the likelihood function with respect to these parameters, subject to any constraints that may be present. The resulting optimal values of the parameters are known as *maximum likelihood estimates*. The method may be applied to functions of discrete or continuous variables.

Aerodynamics. There are many optimization problems concerned with the design, performance and flying qualities of aircraft. The aircraft designer must minimize the structural weight, subject to the structure having sufficient strength and stiffness to carry the critical design loads safely. The cruising altitude should be chosen so as to minimize fuel consumption; it often happens that a steady climb is more economical than flight at constant altitude (explain!).

Aircraft are designed for many different purposes, and in particular cases it may be important to

(a) minimize the take-off run,
(b) maximize the rate of climb,
(c) maximize the ceiling,
(d) maximize the endurance,
(e) minimize the wave drag in supersonic flight.

All these problems are subject to various constraints which, in certain cases, may be so severe that no optimization problem remains.

Chemical engineering. The manager of a chemical plant has to decide on his major objective in running the plant. Should he maximize output? Is this consistent with maximizing profit? To answer these questions requires the solution of at least two optimization problems. The answer to the second question may be 'No', for lower output could mean better quality output, greater efficiency and more valuable by-products.

Operational research. The application of optimization techniques to industrial and commercial problems forms part of the subject of operational research. The fundamental problem of stock control is to choose a stock level and a stock replacement policy which maximize overall profit. The usual assumptions are that losses are incurred if either too much or too little stock is kept. The demand may be known exactly or its frequency function may be assumed. A related problem is that of renewing obsolescent machinery while maintaining maximum efficiency.

Economics. How many new power stations should be built in Britain between now and the year 2000? How many of them should be atomic power stations? These questions lead to very complicated optimization problems; it is not at all clear which quantities should be maximized or minimized and it is even less clear what constraints should be imposed. Nevertheless, problems of this kind obviously need careful study before the crucial decisions are taken.

1.2. STATEMENT OF THE PROBLEM

The problem of maximizing or minimizing a given function

$$z = f(\mathbf{x}), \tag{1.1}$$

subject to the given constraints

$$g_i(\mathbf{x}) \leqslant, = \text{ or } \geqslant b_i, \tag{1.2}$$

is called the *general constrained optimization problem.* The function z appearing in (1.1) is called the *objective function.* In (1.2), the number of independent equality constraints must be less than n, the number of variables, otherwise the problem is overspecified.

Inequalities of $\leqslant$ and $\geqslant$ types can always be converted into equations by introducing *slack* and *surplus variables*, respectively. For example, the inequalities

$$g_1(\mathbf{x}) \leqslant b_1, \quad g_2(\mathbf{x}) \geqslant b_2 \tag{1.3}$$

are respectively equivalent to

$$g_1(\mathbf{x}) + x_{n+1} = b_1, \quad g_2(\mathbf{x}) - x_{n+2} = b_2, \tag{1.4}$$

provided that the slack variable x_{n+1} and the surplus variable x_{n+2} satisfy

$$x_{n+1} \geqslant 0, \quad x_{n+2} \geqslant 0. \tag{1.5}$$

The variables x_j are called *main variables* whenever it is necessary to distinguish them from the slack and surplus variables x_{n+i}. Constraints of the type (1.5) are called *non-negativity restrictions*; in some problems, they are also imposed on the main variables. Although it is perfectly correct to regard non-negativity restrictions as constraints to be included among those of (1.2), it is often found convenient to treat them separately. In general, the effect of substituting (1.4) and (1.5) for (1.3) is to simplify the constraints at the expense of an increased number of variables; this substitution is often extremely useful.

Strict inequality constraints have been omitted from (1.2). This is not a serious limitation in practice, since any constraint of $<$ or $>$ type can be replaced by one of $\leqslant$, $=$ or $\geqslant$ type by means of some simple manipulations. For example, the constraint

$$g_k(\mathbf{x}) < b_k$$

is for all practical purposes equivalent to

$$g_k(\mathbf{x}) \leqslant b_k - \varepsilon,$$

where ε is a suitably small positive constant. The most important reason, however, for restricting the constraints (1.2) to the $\leqslant$, $=$ and $\geqslant$ types is a theoretical one: many fundamental results in optimization theory no longer apply when strict inequality constraints are introduced.

There is no essential difference between a maximizing problem and a minimizing problem, for the values of the x_j which maximize $f(\mathbf{x})$ also minimize $-f(\mathbf{x})$. Thus every maximizing problem can be formulated as a minimizing problem, and vice versa. From now on, problems will be formulated in either the maximizing or the minimizing form, but not both.

Since there is not at present, nor is there ever likely to be, a single recommended method for solving every general constrained optimization problem, it is important to take advantage of any special features that a given problem may possess. It is therefore useful to classify the special cases of the general problem.

The most obvious special case is the *general unconstrained optimization problem*, in which there are no constraints, and the problem is merely (!) to find values of the x_j which maximize $f(\mathbf{x})$. Many modern optimization techniques are designed specifically to solve the general unconstrained optimization problem, for, given a constrained optimization problem, techniques exist which make it possible to write down an equivalent unconstrained problem. Thus the description of an optimization problem as 'unconstrained' is a convenient mathematical classification, but may in fact be a misnomer. Methods for solving unconstrained optimization problems are discussed in Chapters 3 and 4, while methods for dealing with constraints are the subject of Chapter 5.

When every constraint in (1.2) is an equation, we have the *classical optimization problem*:

$$\begin{aligned} &\text{maximize} && z = f(\mathbf{x}), \\ &\text{subject to} && g_i(\mathbf{x}) = b_i. \end{aligned} \quad (1.6)$$

In this problem, the functions f and g_i are assumed to possess continuous first-order partial derivatives with respect to all the variables. Functions with this property are said to belong to the class C_1. Necessary conditions for a maximum can be found by the classical analytic method of Lagrange multipliers; if we assume further that the functions f and g_i possess continuous second-order partial derivatives with respect to all the variables, i.e. if $f, g_i \in C_2$, then sufficient conditions for a maximum can also be found (see Section 1.3). An important advance in optimization theory took place in 1951 when Kuhn and Tucker[51] extended the classical method of Lagrange multipliers to problems with inequality constraints and non-negativity restrictions. Kuhn–Tucker theory is discussed in Chapter 2.

If both $f(\mathbf{x})$ and all the $g_i(\mathbf{x})$ are linear functions of the x_j, we have a *linear programming problem*. Linear programming is still one of the most widely used optimization techniques. There are two principal reasons for this: first, it has many hundreds of useful applications[17,67] and, secondly, extremely large problems can now be solved on electronic computers by means of the simplex method. The simplex method, which was devised by George B. Dantzig in 1947, is an algorithm for the solution of the general linear programming problem. It takes several forms, and the associated theory has been explained in many texts, including Hadley[43] and Walsh,[75] but will not be discussed in detail here.

The function $f(\mathbf{x})$ is said to be *separable* if it is of the form $\sum_j f_j(x_j)$. If both $f(\mathbf{x})$ and all the $g_i(\mathbf{x}_j$ are separable, we have a *separable programming problem*:

$$\begin{aligned} &\text{maximize} && z = \sum_j f_j(x_j), \\ &\text{subject to} && \sum_j g_{ij}(x_j) \leqslant, = \text{ or } \geqslant b_i. \end{aligned} \quad (1.7)$$

Special methods are available for the solution of this problem—see, for example,

Hadley[44], where full accounts will be found of the methods of Charnes and Lemke[18] and Dantzig[21,22]. These methods essentially reduce the separable programming problem to a linear programming problem. Also, general methods tend to be more efficient than usual when they are applied to the separable programming problem, owing to the lack of interaction between the variables.

If either $f(\mathbf{x})$ or one or more of the $g_i(\mathbf{x})$ is nonlinear in any of the variables, we have a *nonlinear programming problem.* Thus every constrained optimization problem defined by (1.1) and (1.2) is either a linear programming problem or a nonlinear programming problem.

If $f(\mathbf{x})$ is a quadratic function of the x_j, while all the $g_i(\mathbf{x})$ are linear in the x_j, we have a *quadratic programming problem.* Many algorithms have been devised for the solution of this type of problem; most of them rely on an extension of the simplex method. The theory of Wolfe's algorithm[78] for the solution of the quadratic programming problem is presented in Section 2.5.

The first step towards choosing an appropriate optimization technique to solve a given problem is to find out whether the problem belongs to any of the special categories mentioned above. Among other factors affecting the choice of method are the time available for a solution, the accuracy required, the computer facilities available, the relative ease with which f, g_i, ∇f, ∇g_i can be evaluated, and whether the variables x_j are continuous or discrete. The significance of these factors will become apparent as various optimization techniques are described.

1.3 CLASSICAL OPTIMIZATION

We now consider the classical optimization problem (1.6) with the object of introducing Lagrange multipliers in the familiar setting of equality constraints. Later, we shall extend their use to the case where inequality constraints are present.

The classical optimization problem (1.6) is:

$$\left.\begin{array}{ll}\text{maximize} & \\ & z = f(\mathbf{x}), \\ \text{subject to} & \\ & g_i(\mathbf{x}) = b_i,\end{array}\right\} \qquad (1.8)$$

where $f, g_i \in C_1$. It is assumed that the constraints are independent.

The basic idea in using Lagrange multipliers is to convert the constrained problem (1.8) into an unconstrained problem. At this point we need some definitions. We say that $f(\mathbf{x})$ has a *constrained local maximum* at $\mathbf{x} = \mathbf{x}^*$ if a number $\varepsilon > 0$ exists such that $f(\mathbf{x}) \leqslant f(\mathbf{x}^*)$ for all $\mathbf{x}$ satisfying both the inequality $|\mathbf{x} - \mathbf{x}^*| < \varepsilon$ and the constraints in (1.8). As an important special case of a constrained local maximum, we say that $f(\mathbf{x})$ has a *constrained global maximum* at $\mathbf{x} = \mathbf{x}^*$ if $f(\mathbf{x}) \leqslant f(\mathbf{x}^*)$ for all $\mathbf{x}$ satisfying the constraints in (1.8). For completeness, we may add that for the unconstrained optimization problem we say simply that $f(\mathbf{x})$ has a *local maximum* or a *global maximum* at $\mathbf{x} = \mathbf{x}^*$

if the above definitions still hold when reference to the constraints in (1.8) is omitted. Any maximum at $\mathbf{x} = \mathbf{x}^*$ is said to be a *strong maximum* if there exists a neighbourhood of $\mathbf{x}^*$ in which no other maximum occurs. Otherwise, the maximum is said to be a *weak maximum.*

Necessary conditions for a constrained local maximum

Assume now that $f(\mathbf{x})$ has a constrained local maximum at $\mathbf{x} = \mathbf{x}^*$. Then, at $\mathbf{x} = \mathbf{x}^*$, we must have

$$\mathrm{d}f \equiv \sum_j \frac{\partial f}{\partial x_j} \mathrm{d}x_j = 0 \tag{1.9}$$

and

$$\mathrm{d}g_i \equiv \sum_j \frac{\partial g_i}{\partial x_j} \mathrm{d}x_j = 0. \tag{1.10}$$

Multiply equations (1.10) by constants $-\lambda_i$ respectively, and add all the resulting equations to equation (1.9). We can then write

$$\mathrm{d}F \equiv \sum_j \frac{\partial F}{\partial x_j} \mathrm{d}x_j = 0, \tag{1.11}$$

where the *Lagrangian function* F is defined by

$$F(\mathbf{x},\boldsymbol{\lambda}) \equiv f(\mathbf{x}) + \sum_i \lambda_i [b_i - g_i(\mathbf{x})] \tag{1.12}$$

and the *Lagrange multipliers* λ_i are the components of the m-vector $\boldsymbol{\lambda}$. Because of the m independent constraints in (1.8), only $(n-m)$ of the variables x_j are independent. Let these be $x_{m+1},\ldots,x_n$. Now choose the λ_i, which so far are arbitrary, in such a way that

$$\frac{\partial F}{\partial x_i} = 0. \tag{1.13}$$

Then equation (1.11) becomes

$$\frac{\partial F}{\partial x_{m+1}} dx_{m+1} + \;\ldots\; + \frac{\partial F}{\partial x_n} dx_n = 0. \tag{1.14}$$

But the variables $x_{m+1},\ldots,x_n$ are independent and, by allowing each of them in turn to vary, we obtain, from equation (1.14),

$$\frac{\partial F}{\partial x_{m+1}} = \;\ldots\; = \frac{\partial F}{\partial x_n} = 0. \tag{1.15}$$

Thus, from equations (1.13) and (1.15), we have

$$\frac{\partial F}{\partial x_j} \equiv \frac{\partial f}{\partial x_j} - \sum_i \lambda_i \frac{\partial g_i}{\partial x_j} = 0, \tag{1.16}$$

which are n necessary conditions for the existence of a constrained local maximum of $f(\mathbf{x})$.

Hessian

Conditions (1.16) also apply when $f(\mathbf{x})$ is to be *minimized* subject to the constraints of problem (1.8). More generally, any point satisfying equations (1.16) is called a *stationary point* or *critical point* of $f(\mathbf{x})$. The words 'constrained', 'unconstrained', 'local' and 'global' may be added to these and other definitions, as appropriate. A stationary point at which $f(\mathbf{x})$ takes on a maximum or minimum value is called an *extremum* or *turning point*, and the corresponding value of the function is called an *extreme value* or *turning value.* Note that not all maxima and minima occur at stationary points; on the other hand, not all stationary points correspond to maxima or minima. An *optimal* point $\mathbf{x}^*$ is any point at which $f(\mathbf{x})$ takes on a maximum or minimum value; an asterisk will always denote an optimal value.

Assuming the existence of the Lagrange multipliers in equations (1.16), the implicit function theorem[19] guarantees that these equations, together with the constraint equations in (1.8), can be solved for the $(m+n)$ unknowns λ_i, x_j to give $\lambda = \lambda^*$, $\mathbf{x} = \mathbf{x}^*$. On page 13 we return to the question of the existence of the Lagrange multipliers. It is interesting to note that the necessary conditions of (1.16) and (1.8) can be expressed in the symmetrical form

$$\frac{\partial F}{\partial x_j} = 0, \qquad \frac{\partial F}{\partial \lambda_i} = 0. \tag{1.17}$$

Sufficient conditions for a constrained local maximum

It is important to remember that the conditions (1.17) are not sufficient for the existence of a constrained local maximum of $f(\mathbf{x})$. Sufficient conditions involve second- or higher-order derivatives of $f(\mathbf{x})$, as in the unconstrained optimization problem. A full discussion of these somewhat complicated conditions would lead us too far afield. Instead, we shall outline two methods for obtaining sufficient conditions, illustrating their use by means of an example. For further details, the interested reader is referred to Hancock.[45]

We assume that $f, g_i \in C_2$. A sufficient condition for the existence of a constrained local maximum of the function f is that $\mathrm{d}^2 f < 0$ at the optimal point, subject to $\mathrm{d}g_i = 0$ and $\mathrm{d}^2 g_i = 0$. Hence, taking the differential of the left-hand side of equation (1.9), we find

$$\mathrm{d}(\mathrm{d}f) = \mathrm{d}^2 f = \mathbf{dx}'\mathbf{H}_f(\mathbf{x}^*)\mathbf{dx} + [\nabla f(\mathbf{x}^*)]\,\mathbf{d}^2\mathbf{x} < 0, \tag{1.18}$$

where

$$\mathbf{H}_f(\mathbf{x}) \equiv \left\{\frac{\partial^2 f}{\partial x_j \partial x_k}\right\} \quad (j,k = 1,\ldots,n)$$

is the Hessian matrix of $f(\mathbf{x})$. In (1.18), the vector $\mathbf{dx} \equiv [\mathrm{d}x_1,\ldots,\mathrm{d}x_n]$ must satisfy the vector-matrix form of equations (1.10), namely

$$\mathrm{d}g_i \equiv [\nabla g_i(\mathbf{x}^*)]'\,\mathbf{dx} = 0, \tag{1.19}$$

and must also satisfy

$$\mathrm{d}^2 g_i \equiv \mathbf{dx}'\mathbf{H}g_i(\mathbf{x}^*)\mathbf{dx} + [\nabla g_i(\mathbf{x}^*)]'\mathbf{d}^2\mathbf{x} = 0. \tag{1.20}$$

As in the theory of unconstrained optimization, condition (1.18) is not necessary for the existence of a local maximum. If $\mathrm{d}^2 f = 0$, then higher-order derivatives must be considered in order to determine whether or not a local maximum of $f(\mathbf{x})$ has been found, although geometry may be used in simple cases.

An alternative sufficient condition for a constrained local maximum of $f(\mathbf{x})$ involves the Hessian matrix $\mathbf{H}_F(\mathbf{x}, \lambda)$ of the Lagrangian function F. Let $\mathbf{x}^*$, λ^* denote the optimal values of $\mathbf{x}, \lambda$, respectively, and expand $F(\mathbf{x}, \lambda^*)$ by the second mean value theorem in a small neighbourhood of $\mathbf{x}^*$:

$$F(\mathbf{x}, \lambda^*) = F(\mathbf{x}^*, \lambda^*) + \mathbf{h}'\nabla F(\mathbf{x}^*, \lambda^*) + \tfrac{1}{2}\mathbf{h}'\mathbf{H}_F(\bar{\mathbf{x}}_0, \lambda^*)\mathbf{h}, \tag{1.21}$$

where

$$\mathbf{h} = \mathbf{x} - \mathbf{x}^*,$$

$$\bar{\mathbf{x}}_0 = \mathbf{x}^* + \theta_0\mathbf{h} \quad (0 \leqslant \theta_0 \leqslant 1).$$

Now assume that $\mathbf{x} = \mathbf{x}^* + \mathbf{h}$ satisfies the constraints in (1.8). Then, from (1.12), we have

$$F(\mathbf{x}, \lambda^*) = f(\mathbf{x}),$$

and equation (1.21) becomes

$$\begin{aligned} f(\mathbf{x}^* + \mathbf{h}) &= f(\mathbf{x}^*) + \mathbf{h}'\Big[\nabla f(\mathbf{x}^*) - \sum_i \lambda_i^* \nabla g_i(\mathbf{x}^*)\Big] + \tfrac{1}{2}\mathbf{h}'\mathbf{H}_F(\bar{\mathbf{x}}_0, \lambda^*)\mathbf{h} \\ &= f(\mathbf{x}^*) + \tfrac{1}{2}\mathbf{h}'\mathbf{H}_F(\bar{\mathbf{x}}_0, \lambda^*)\mathbf{h}, \end{aligned} \tag{1.22}$$

where we have used equations (1.16). Also,

$$g_i(\mathbf{x}^* + \mathbf{h}) = g_i(\mathbf{x}^*) + \mathbf{h}'\nabla g_i(\bar{\mathbf{x}}_i),$$

where

$$\bar{\mathbf{x}}_i = \mathbf{x}^* + \theta_i\mathbf{h} \qquad (0 \leqslant \theta_i \leqslant 1).$$

This gives

$$b_i = b_i + \mathbf{h}'\nabla g_i(\bar{\mathbf{x}}_i),$$

or

$$\mathbf{h}'\nabla g_i(\bar{\mathbf{x}}_i) = 0. \tag{1.23}$$

Thus, from equations (1.22) and (1.23), a sufficient condition for the existence of a constrained local maximum of $f(\mathbf{x})$ at $\mathbf{x} = \mathbf{x}^*$ is that

$$\mathbf{h}'\mathbf{H}_F(\bar{\mathbf{x}}_0, \lambda^*)\mathbf{h} < 0$$

for all $\mathbf{h}$ satisfying equation (1.23), with $|\mathbf{h}|$ sufficiently small for the definition of a constrained local maximum (page 6) to apply. Since the elements of $\mathbf{H}_F(\mathbf{x}, \lambda)$ and $\nabla g_i(\mathbf{x})$ are continuous functions of $\mathbf{x}$, the required sufficient condition can be stated as follows:

$$\left.\begin{array}{l}\text{the quadratic form } \mathbf{h}'\mathbf{H}_F(\mathbf{x}^*, \lambda^*)\mathbf{h} \text{ is negative definite (positive} \\ \text{definite for a minimizing problem) for all } \mathbf{h} \text{ satisfying} \\ \qquad\qquad \mathbf{h}'\nabla g_i(\mathbf{x}^*) = 0.\end{array}\right\} \tag{1.24}$$

The sufficient condition (1.24) is not in a form that can be used immediately. However, we shall now prove that it is equivalent to the following condition:

$$
\left.\begin{array}{l}
\text{every root of the polynomial equation} \\[1ex]
\qquad P(\mu) \equiv \left|\begin{array}{c:c} \mathbf{A}-\mu\mathbf{I} & \mathbf{D} \\ \hdashline \mathbf{D}' & \mathbf{O} \end{array}\right| = 0 \\[1ex]
\text{is strictly negative (positive for a minimizing problem), where} \\[1ex]
\qquad \mathbf{A} = \mathbf{H}_F(\mathbf{x}^*, \boldsymbol{\lambda}^*) \\[1ex]
\text{and the } i\text{th column of } \mathbf{D} \text{ is} \\[1ex]
\qquad \mathbf{d}_i = \nabla g_i(\mathbf{x}^*).
\end{array}\right\} \qquad (1.25)
$$

We can assume without loss of generality that the vectors **h** are unit vectors. Then the condition (1.24) is equivalent to the statement that the maximum value of $\mathbf{h}'\mathbf{A}\mathbf{h}$, subject to the constraints

$$\mathbf{h}'\mathbf{h} = 1, \qquad \mathbf{h}'\mathbf{d}_i = 0, \qquad (1.26)$$

is strictly negative.

Define the Lagrangian function

$$\mathscr{F}(\mathbf{h},\mu_0,\boldsymbol{\mu}) \equiv \mathbf{h}'\mathbf{A}\mathbf{h} + \mu_0(1-\mathbf{h}'\mathbf{h}) - \sum_i \mu_i \mathbf{h}'\mathbf{d}_i,$$

where μ_0 and the μ_i are Lagrange multipliers and $\boldsymbol{\mu} \equiv [\mu_1, \ldots, \mu_m]$. It is necessary for a constrained maximum of $\mathbf{h}'\mathbf{A}\mathbf{h}$ that μ_0, μ_i exist such that $\nabla_h \mathscr{F}(\mathbf{h},\mu_0,\boldsymbol{\mu}) = 0$, i.e.

$$2\mathbf{A}\mathbf{h} - 2\mu_0\mathbf{h} - \sum_i \mu_i \mathbf{d}_i = 0. \qquad (1.27)$$

Also, since the function $\mathbf{h}'\mathbf{A}\mathbf{h}$ is continuous on the closed, bounded set defined by the constraints (1.26), it has a maximum on this set. Premultiplying equation (1.27) by $\mathbf{h}'$ and taking account of the constraints (1.26), we obtain

$$\mathbf{h}'\mathbf{A}\mathbf{h} = \mu_0,$$

and this is the constrained maximum value of $\mathbf{h}'\mathbf{A}\mathbf{h}$. This value must be strictly negative, i.e. we require $\mu_0 < 0$.

Now equation (1.27) may be written

$$2(\mathbf{A}-\mu_0\mathbf{I})\mathbf{h} - \mathbf{D}\boldsymbol{\mu} = 0.$$

Also, from the second of the constraint equations (1.26),

$$\mathbf{D}'\mathbf{h} = 0.$$

Hence, by the theory of linear algebraic equations, non-trivial **h** and $\boldsymbol{\mu}$ exist if and only if

$$\left|\begin{array}{c:c} 2(\mathbf{A}-\mu_0\mathbf{I}) & -\mathbf{D} \\ \hdashline \mathbf{D}' & \mathbf{O} \end{array}\right| = 0,$$

i.e. if and only if μ_0 is a root of

$$\left|\begin{array}{c:c} \mathbf{A}-\mu\mathbf{I} & \mathbf{D} \\ \hdashline \mathbf{D}' & \mathbf{O} \end{array}\right| = 0. \tag{1.28}$$

The requirement $\mu_0 < 0$ implies that every root of equation (1.28) must be strictly negative. This completes the proof of the equivalence of conditions (1.24) and (1.25).

Incidentally, since **A** is the matrix of a quadratic form, it can be assumed to be symmetric. Then the determinant

$$\left|\begin{array}{c:c} \mathbf{A} & \mathbf{D} \\ \hdashline \mathbf{D}' & \mathbf{O} \end{array}\right|$$

is symmetric, and hence every root of $P(\mu) = 0$ is real. The proof is left to the exercises.

Exercise

What happens to equation (1.28) when $\mathbf{D} = \mathbf{O}$?

Example 1.1

Find the maximum value of $\sin A \sin B \sin C$, *given that* $A+B+C=\pi$.

Solution

Define the Lagrangian function

$$F(A,B,C,\lambda) \equiv \sin A \sin B \sin C + \lambda(\pi - A - B - C).$$

Using equations (1.16), we find

$$\left.\begin{aligned} \frac{\partial F}{\partial A} &\equiv \cos A \sin B \sin C - \lambda = 0, \\ \frac{\partial F}{\partial B} &\equiv \sin A \cos B \sin C - \lambda = 0, \\ \frac{\partial F}{\partial C} &\equiv \sin A \sin B \cos C - \lambda = 0. \end{aligned}\right\} \tag{1.29}$$

Equations (1.29), together with the constraint equation

$$A+B+C=\pi, \tag{1.30}$$

are necessary conditions for a maximum. Eliminating λ between the first two equations of (1.29), we obtain

$$(\sin A \cos B - \cos A \sin B)\sin C = 0,$$

i.e.

$$\sin(A-B) = 0 \qquad or \qquad \sin C = 0. \tag{1.31}$$

Similarly,

$$\sin(B-C) = 0 \qquad or \qquad \sin A = 0. \tag{1.32}$$

Let m_1, m_2, n_1, n_2 be arbitrary integers. Then, from equations (1.30) to (1.32), we find that there are four possible sets of solutions for A, B and C, given by

$$\text{(i)}\quad \left.\begin{aligned} A+B+C&=\pi,\\ A-B&=m_1\pi,\\ B-C&=m_2\pi;\end{aligned}\right\} \tag{1.33}$$

$$\text{(ii)}\quad \left.\begin{aligned} A+B+C&=\pi,\\ C&=n_1\pi,\\ B-C&=m_2\pi;\end{aligned}\right\} \tag{1.34}$$

$$\text{(iii)}\quad \left.\begin{aligned} A+B+C&=\pi,\\ A-B&=m_1\pi,\\ A&=n_2\pi;\end{aligned}\right\} \tag{1.35}$$

$$\text{(iv)}\quad \left.\begin{aligned} A+B+C&=\pi,\\ C&=n_1\pi,\\ A&=n_2\pi.\end{aligned}\right\} \tag{1.36}$$

Cases (ii), (iii) and (iv) may be ignored, since they give a value of zero for $\sin A \sin B \sin C$, and this is obviously not its maximum value. The solution of equations (1.33), case (i), is

$$\left.\begin{aligned} A&=(1+2m_1+m_2)(\pi/3),\\ B&=(1-m_1+m_2)(\pi/3),\\ C&=(1-m_1-2m_2)(\pi/3).\end{aligned}\right\} \tag{1.37}$$

Hence $\sin A$, $\sin B$ and $\sin C$ can each take the values $\pm\sqrt{3}/2$ or 0. The constrained global maximum value of $\sin A \sin B \sin C$ is therefore $3\sqrt{3}/8$ and occurs when $\sin A = \sin B = \sin C = \sqrt{3}/2$. Taking account of equations (1.37) or, more simply, using equation (1.30), we find that

$$A^*, B^*, C^* = (2k_1+\tfrac{1}{3})\pi,\ (2k_2+\tfrac{1}{3})\pi,\ -(2k_1+2k_2-\tfrac{1}{3})\pi,$$

where k_1 and k_2 are any integers.

It should be noted that we have not assumed that A, B and C are necessarily positive; this condition would involve inequality constraints in the form of non-negativity restrictions.

Although in the present example the choice of the optimal solution from candidates (i) to (iv) was obvious, it is instructive to verify analytically that a true maximum has been found. For purposes of illustration, we shall carry out this procedure twice, using the alternative sufficiency conditions (1.18) and (1.25).

First, for the method of (1.18), we have

$$f = \sin A \sin B \sin C,$$

and so

$$\begin{aligned} \mathrm{d}f &= \frac{\partial f}{\partial A}\mathrm{d}A + \frac{\partial f}{\partial B}\mathrm{d}B + \frac{\partial f}{\partial C}\mathrm{d}C\\ &= \cos A \sin B \sin C\, \mathrm{d}A + \sin A \cos B \sin C\, \mathrm{d}B + \sin A \sin B \cos C\, \mathrm{d}C\\ &= 0,\end{aligned}$$

where, in order to satisfy the constraint (1.30), the differentials dA, dB, dC must satisfy

$$\mathrm{d}A + \mathrm{d}B + \mathrm{d}C = 0,$$

which corresponds to equation (1.19). Differentiating again, we obtain

$$\mathrm{d}^2 f = \left[\frac{\partial^2 f}{\partial A^2}(\mathrm{d}A)^2 + \frac{\partial f}{\partial A}\mathrm{d}^2 A + \frac{\partial^2 f}{\partial A\,\partial B}\mathrm{d}A\,\mathrm{d}B + \frac{\partial^2 f}{\partial A\,\partial C}\mathrm{d}A\,\mathrm{d}C\right] + \dots + \dots$$

and

$$\mathrm{d}^2 A + \mathrm{d}^2 B + \mathrm{d}^2 C = 0, \tag{1.38}$$

which correspond to equations (1.18) and (1.20), respectively. Hence

$$\begin{aligned}\mathrm{d}^2 f = [&-\sin A \sin B \sin C(\mathrm{d}A)^2 + \cos A \sin B \sin C\,\mathrm{d}^2 A \\ &+ \cos A \cos B \sin C\,\mathrm{d}A\,\mathrm{d}B + \cos A \sin B \cos C\,\mathrm{d}A\,\mathrm{d}C] + \dots + \dots\end{aligned}$$

At the optimal point, this gives

$$\begin{aligned}\mathrm{d}^2 f &= [-\tfrac{3\sqrt{3}}{8}(\mathrm{d}A)^2 + \tfrac{3}{8}\mathrm{d}^2 A + \tfrac{\sqrt{3}}{8}\mathrm{d}A\,\mathrm{d}B + \tfrac{\sqrt{3}}{8}\mathrm{d}A\,\mathrm{d}C] + \dots + \dots \\ &= [-\tfrac{3\sqrt{3}}{8}(\mathrm{d}A)^2 + \tfrac{3}{8}\mathrm{d}^2 A - \tfrac{\sqrt{3}}{8}(\mathrm{d}A)^2] + \dots + \dots \\ &= -\tfrac{\sqrt{3}}{2}[(\mathrm{d}A)^2 + (\mathrm{d}B)^2 + (\mathrm{d}C)^2] + \tfrac{3}{8}(\mathrm{d}^2 A + \mathrm{d}^2 B + \mathrm{d}^2 C) \\ &< 0,\end{aligned}$$

where we have used equation (1.38). The above inequality is the required result.

Using the alternative method, which is based on condition (1.25), we find

$$\mathbf{H}_F(A,B,C) = \begin{pmatrix} -\sin A \sin B \sin C & \cos A \cos B \sin C & \cos A \sin B \cos C \\ \cos A \cos B \sin C & -\sin A \sin B \sin C & \sin A \cos B \cos C \\ \cos A \sin B \cos C & \sin A \cos B \cos C & -\sin A \sin B \sin C \end{pmatrix},$$

$$\mathbf{D} = \begin{pmatrix} 1 \\ 1 \\ 1 \end{pmatrix}.$$

We have to show that every root of $P(\mu) = 0$ is strictly negative, where

$$P(\mu) \equiv \begin{vmatrix} (-\frac{3\sqrt{3}}{8} - \mu) & \frac{\sqrt{3}}{8} & \frac{\sqrt{3}}{8} & 1 \\ \frac{\sqrt{3}}{8} & (-\frac{3\sqrt{3}}{8} - \mu) & \frac{\sqrt{3}}{8} & 1 \\ \frac{\sqrt{3}}{8} & \frac{\sqrt{3}}{8} & (-\frac{3\sqrt{3}}{8} - \mu) & 1 \\ 1 & 1 & 1 & 0 \end{vmatrix}.$$

The equation $P(\mu) = 0$ reduces to

$$(\mu + \tfrac{\sqrt{3}}{2})^2 = 0,$$

as students of determinants will see at a glance. The double root of this equation is strictly negative, as required.

On page 8 we assumed the existence of the Lagrange multipliers in equations (1.16). The question of existence is not trivial for, unless the λ_i exist, the necessary conditions (1.16) for a maximum are meaningless; see Young[80] for a thorough discussion of this topic in a more general setting.

From the theory of systems of linear algebraic equations, a necessary and sufficient condition for the existence of the λ_i in (1.16) is that the $(n \times m)$ matrix

$$\mathbf{G} = \left\{\frac{\partial g_i}{\partial x_j}\right\} \tag{1.39}$$

should have the same rank as the $(n \times (m+1))$ matrix

$$(\mathbf{G} \vdots \nabla f) = \left\{\frac{\partial g_i}{\partial x_j} \vdots \frac{\partial f}{\partial x_j}\right\}. \tag{1.40}$$

Furthermore, the λ_i are unique if and only if the common rank of these matrices is m.

We now consider a slightly more general form of the theory of classical optimization in which the Lagrangian function is defined by

$$F(\mathbf{x}, \lambda_0, \boldsymbol{\lambda}) \equiv \lambda_0 f(\mathbf{x}) + \sum_i \lambda_i [b_i - g_i(\mathbf{x})].$$

The case we have considered so far corresponds to $\lambda_0 = 1$. Equations (1.16) now become

$$\frac{\partial F}{\partial x_j} \equiv \lambda_0 \frac{\partial f}{\partial x_j} - \sum_i \lambda_i \frac{\partial g_i}{\partial x_j} = 0. \tag{1.41}$$

Equations (1.41), like equations (1.16), are necessary conditions for the existence of a constrained local maximum of $f(\mathbf{x})$.

Suppose that $\text{rank}(\mathbf{G} \vdots \nabla f) > \text{rank}(\mathbf{G})$; this implies $\text{rank}(\mathbf{G} \vdots \nabla f) = \text{rank}(\mathbf{G}) + 1$. In this case, we have just seen that the λ_i in (1.41) do not exist when $\lambda_0 = 1$. However, if we set $\lambda_0 = 0$, and if $\text{rank}(\mathbf{G}) \leqslant m - 1$, then non-trivial λ_i satisfying (1.41) do exist, but are not unique.

Finally, since equations (1.41) are homogeneous in $\lambda_0, \ldots, \lambda_m$, we can assume without loss of generality that $\lambda_0 = 0$ or 1. The case $\lambda_0 = 0$ is often referred to as the *abnormal case*; it is illustrated in the following example.

Maximize

$$f(\mathbf{x}) \equiv x_1 + x_2 + x_3,$$

subject to

$$g_1(\mathbf{x}) \equiv x_1^2 + x_2^2 + x_3^2 = b.$$

Solution

The Lagrangian function is

$$F(\mathbf{x}, \lambda_0, \boldsymbol{\lambda}) \equiv \lambda_0(x_1 + x_2 + x_3) + \lambda_1(b - x_1^2 - x_2^2 - x_3^2).$$

Equations (1.41) give

$$\left.\begin{aligned} \lambda_0 - 2x_1\lambda_1 &= 0, \\ \lambda_0 - 2x_2\lambda_1 &= 0, \\ \lambda_0 - 2x_3\lambda_1 &= 0. \end{aligned}\right\} \tag{1.42}$$

Also,

$$\mathbf{G} = \begin{pmatrix} 2x_1 \\ 2x_2 \\ 2x_3 \end{pmatrix}, \qquad \nabla f = \begin{pmatrix} 1 \\ 1 \\ 1 \end{pmatrix}.$$

There are two cases to consider.

(i) $b > 0$. From equations (1.42), we have

$$x_1 = x_2 = x_3 = \lambda_0/2\lambda_1 \neq 0, \tag{1.43}$$

using the fact that $b \neq 0$. We therefore take $\lambda_0 = 1$. Hence the optimal solution is

$$x_1{}^* = x_2{}^* = x_3{}^* = (b/3)^{1/2}; \quad f(\mathbf{x}^*) = (3b)^{1/2}, \text{ with } \lambda_1{}^* = (3/4b)^{1/2}.$$

Note that $\text{rank}(\mathbf{G} \vdots \nabla f) = \text{rank}(\mathbf{G}) = 1$.

(ii) $b = 0$. Instead of (1.43), we now have

$$x_1 = x_2 = x_3 = \lambda_0/2\lambda_1 = 0.$$

Hence $\lambda_0 = 0$; and $\lambda_1 \neq 0$, but is otherwise arbitrary. The optimal solution is

$$x_1{}^* = x_2{}^* = x_3{}^* = 0; \quad f(\mathbf{x}^*) = 0.$$

Note that $\text{rank}(\mathbf{G} \vdots \nabla f) = 1$ and $\text{rank}(\mathbf{G}) = 0$.

From now on, unless otherwise stated, we shall assume that $\lambda_0 = 1$.

So far, Lagrange multipliers have been presented as convenient constants which are brought into use to assist in the solution of the classical optimization problem. However, they have a much more interesting interpretation.

Let $\mathbf{x}^*$ and $z^* = f(\mathbf{x}^*)$ be the optimal values of $\mathbf{x}$ and z for problem (1.8), and consider the slightly more general situation in which the b_i are allowed to vary. The question arises as to how z^* changes when one or more of the b_i are changed. We have

$$\frac{\partial f(\mathbf{x}^*)}{\partial b_i} = \sum_j \frac{\partial f}{\partial x_j{}^*} \frac{\partial x_j{}^*}{\partial b_i} \tag{1.44}$$

and

$$\frac{\mathrm{d}g_k(\mathbf{x}^*)}{\mathrm{d}b_i} = \delta_{ik} = \sum_j \frac{\partial g_k}{\partial x_j{}^*} \frac{\partial x_j{}^*}{\partial b_i},$$

where δ_{ik} is the Kronecker symbol. Multiplying the last equation by the Lagrange multiplier $\lambda_k{}^*$ and summing over k, we obtain

$$\sum_{k=1}^{m} \lambda_k{}^* \delta_{ik} = \sum_{k=1}^{m} \lambda_k{}^* \sum_j \frac{\partial g_k}{\partial x_j{}^*} \frac{\partial x_j{}^*}{\partial b_i}. \tag{1.45}$$

Because of equation (1.45), we can write equation (1.44) as

$$\frac{\partial f(\mathbf{x}^*)}{\partial b_i} = \sum_{k=1}^{m} \lambda_k{}^* \delta_{ik} + \sum_j \frac{\partial f}{\partial x_j{}^*} \frac{\partial x_j{}^*}{\partial b_i} - \sum_{k=1}^{m} \lambda_k{}^* \sum_j \frac{\partial g_k}{\partial x_j{}^*} \frac{\partial x_j{}^*}{\partial b_i}$$

$$= \sum_{k=1}^{m} \lambda_k{}^*\delta_{ik} + \sum_j \left[\frac{\partial f}{\partial x_j{}^*} - \sum_{k=1}^{m} \lambda_k{}^* \frac{\partial g_k}{\partial x_j{}^*}\right] \frac{\partial x_j{}^*}{\partial b_i}.$$

The expression in square brackets vanishes for each value of j because of equations (1.16). Hence we obtain the important result

$$\frac{\partial z^*}{\partial b_i} = \lambda_i{}^*. \tag{1.46}$$

The Lagrange multiplier $\lambda_i{}^*$ therefore measures the rate of increase of z^* with respect to b_i, due regard being paid to signs. In this sense, $\lambda_i{}^*$ indicates the relative importance of the ith constraint. In economics, where z^* often represents an optimal cost and b_i the amount of some available resource, $\lambda_i{}^*$ is called a *shadow cost*. This economic interpretation of Lagrange multipliers can be extended to the case where inequality constraints are present. In particular, Lagrange multipliers may be identified with the dual variables in a linear programming problem; extensive use is made of this relationship in linear economic analysis.[3, 23, 38]

1.4 CLASSICAL TREATMENT OF INEQUALITY CONSTRAINTS

The classical theory of the previous section can be extended to include inequality constraints by adding the necessary slack and surplus variables to these constraints.

Suppose the given problem is:

maximize

$$\left.\begin{aligned} &z = f(\mathbf{x}), \\ &\text{subject to} \\ &g_i(\mathbf{x}) \leqslant b_i \qquad (i=1,\ldots.,u), \\ &g_i(\mathbf{x}) \geqslant b_i \qquad (i=u+1,\ldots.,v), \\ &g_i(\mathbf{x}) = b_i \qquad (i=v+1,\ldots.,m), \\ &\mathbf{x} \geqslant 0. \end{aligned}\right\} \tag{1.47}$$

Add slack and surplus variables x_{n+i} $(i=1,\ldots.,v)$ to the inequality constraints:

$$\left.\begin{aligned} g_i(\mathbf{x}) + x_{n+i} &= b_i \qquad (i=1,\ldots.,u), \\ g_i(\mathbf{x}) - x_{n+i} &= b_i \qquad (i=u+1,\ldots.,v), \\ x_{n+i} &\geqslant 0 \qquad (i=1,\ldots.,v). \end{aligned}\right\} \tag{1.48}$$

Notice that the x_{n+i} $(i=1,\ldots.,v)$ satisfy non-negativity restrictions, as in linear programming. The non-negativity restrictions on the main variables in problem (1.47) are included for completeness; they are not always present in nonlinear programming problems.

For each non-negativity restriction $x_{n+i} \geqslant 0$ $(i=1,....,v)$, there are two cases to consider: either $x_{n+i}^* = 0$ or $x_{n+i}^* > 0$. In the first case, the optimal point $\mathbf{x}^*$ lies on the boundary $g_i(\mathbf{x}) = b_i$ of the corresponding constraint and we say that this constraint is *active*. In the second case, the optimal point does not lie on the constraint boundary and we say that the constraint is *passive* or *inactive*. For notational reasons we shall use the term 'passive'. If we knew *a priori* which constraints were active and which passive, we could dispense with inequality constraints altogether; the former would be written as equality constraints and the latter ignored. However, we can determine which constraints are active, and thus solve problem (1.47), by the following systematic trial and error method.

1. Assume that every inequality constraint is passive, i.e. ignore all the inequality constraints, and find the *global* maximum of $f(\mathbf{x})$ subject to the equality constraints and the non-negativity restrictions $\mathbf{x} \geqslant \mathbf{0}$. This is a classical optimization problem in the interior of the feasible region $(\mathbf{x} > \mathbf{0})$, but also involves a search on the boundaries of the feasible region ($x_j = 0$ for at least one value of j). If the point obtained by this procedure also satisfies all the inequality constraints, then the problem is solved. If not,
2. Treat one inequality constraint as active and repeat step 1.
3. Repeat step 2 until every inequality constraint has been tried in turn.
4. Treat two inequality constraints as active and repeat step 1.
5. Repeat step 4 until every pair of inequality constraints has been tried in turn.
6. Treat three inequality constraints as active, etc., continuing until the point so obtained satisfies the remaining inequality constraints.

An obvious disadvantage of this method is that it can be time-consuming when there are several inequality constraints and non-negativity restrictions. With m inequality constraints and taking the worst case, 2^m classical optimization problems have to be solved in the interior of the feasible region. Also, in searching the boundaries of the feasible region, it is necessary to consider separately the cases in which each variable in turn is set equal to zero, each pair of variables in turn is set equal to zero, etc., provided always that the number of variables set equal to zero is not so large that the problem is over-specified.

There is an interesting, though rather obvious, relationship between passive constraints and the corresponding Lagrange multipliers. Suppose we construct a Lagrangian function $F(\mathbf{x},\mathbf{x}_s,\lambda)$ for problem (1.47) with the inequality constraints represented as in (1.48), where $\mathbf{x}_s$ is the vector of slack and surplus variables. Then, for any $x_{n+i} > 0$ $(i=1,....,v)$, a necessary condition for a constrained local maximum of $f(\mathbf{x})$ is $\partial F/\partial x_{n+i} = 0$ at the optimal point, and this gives $\lambda_i^* = 0$.

Thus, if a constraint is passive, the corresponding Lagrange multiplier takes the value zero at the optimal point. This result merely confirms the fact that passive constraints can be ignored.

Example 1.3

Maximize

$$f(\mathbf{x}) \equiv x_1 + 2x_2 + 3x_3,$$

subject to

$$x_1 + x_2 + x_3 \leqslant 5, \tag{1.49}$$

$$x_1{}^2 + x_2{}^2 + x_3{}^2 \leqslant 20, \tag{1.50}$$

$$x_1, x_2, x_3 \geqslant 0.$$

Solution

1. If both inequality constraints are ignored, then $f(\mathbf{x})$ is unbounded and the constraints are obviously violated.

2. Suppose that the constraint (1.49) is active. Since the resulting problem is a linear programming problem, there is no need to introduce a Lagrangian function nor to conduct a separate search on the boundaries of the feasible region. The optimal solution is obviously

$$x_1{}^* = 0, \ x_2{}^* = 0, \ x_3{}^* = 5; \ f(\mathbf{x}^*) = 15.$$

However, this solution does not satisfy the constraint (1.50).

3. Suppose that the constraint (1.50) is active. Define

$$F(\mathbf{x}, \lambda) \equiv x_1 + 2x_2 + 3x_3 + \lambda(20 - x_1{}^2 - x_2{}^2 - x_3{}^2). \tag{1.51}$$

Then

$$\left.\begin{aligned} \frac{\partial F}{\partial x_1} &\equiv 1 - 2\lambda x_1 = 0, \\ \frac{\partial F}{\partial x_2} &\equiv 2 - 2\lambda x_2 = 0, \\ \frac{\partial F}{\partial x_3} &\equiv 3 - 2\lambda x_3 = 0. \end{aligned}\right\} \tag{1.52}$$

Solving equations (1.52) together with the constraint *equation* (1.50), we find

$$x_1{}^* = \sqrt{\frac{10}{7}}, \ x_2{}^* = \sqrt{\frac{40}{7}}, \ x_3{}^* = \sqrt{\frac{90}{7}}; \ f(\mathbf{x}^*) = 16{\cdot}73. \tag{1.53}$$

Next, we search along the boundary $x_1 = 0$. The Lagrangian function (1.51) is replaced by

$$F(\mathbf{x}, \lambda) \equiv 2x_2 + 3x_3 + \lambda(20 - x_2{}^2 - x_3{}^2).$$

Then

$$\left.\begin{aligned} \frac{\partial F}{\partial x_2} &\equiv 2 - 2\lambda x_2 = 0, \\ \frac{\partial F}{\partial x_3} &\equiv 3 - 2\lambda x_3 = 0. \end{aligned}\right\} \tag{1.54}$$

Solving equations (1.54) together with the constraint

$$x_2^2 + x_3^2 = 20,$$

we find

$$x_2^* = \sqrt{\frac{80}{13}},\ x_3^* = \sqrt{\frac{180}{13}};\ f(\mathbf{x}^*) = 16{\cdot}13.$$

Similarly, at the optimal point on the boundary $x_2 = 0$,

$$x_1^* = \sqrt{2},\ x_3^* = 3\sqrt{2};\ f(\mathbf{x}^*) = 14{\cdot}14,$$

and on the boundary $x_3 = 0$,

$$x_1^* = 2,\ x_2^* = 4;\ f(\mathbf{x}^*) = 10.$$

On the boundaries $\{x_2 = x_3 = 0\}$, $\{x_3 = x_1 = 0\}$, $\{x_1 = x_2 = 0\}$, we obtain, respectively,

$$x_1^* = \sqrt{20},\ x_2^* = \sqrt{20},\ x_3^* = \sqrt{20};\ f(\mathbf{x}^*) = \sqrt{20},\ 2\sqrt{20},\ 3\sqrt{20} = 13{\cdot}42.$$

Thus the global maximum of $f(\mathbf{x})$ for this sub-problem is given by (1.53). However, this solution does not satisfy the constraint (1.49), and so we proceed to the case in which both inequality constraints are active.

4. Suppose that the constraints (1.49) and (1.50) are both active. Define

$$F(\mathbf{x}, \lambda) \equiv x_1 + 2x_2 + 3x_3 + \lambda_1(5 - x_1 - x_2 - x_3) + \lambda_2(20 - x_1^2 - x_2^2 - x_3^2).$$

Then

$$\left.\begin{aligned} \frac{\partial F}{\partial x_1} &\equiv 1 - \lambda_1 - 2\lambda_2 x_1 = 0, \\ \frac{\partial F}{\partial x_2} &\equiv 2 - \lambda_1 - 2\lambda_2 x_2 = 0, \\ \frac{\partial F}{\partial x_3} &\equiv 3 - \lambda_1 - 2\lambda_2 x_3 = 0. \end{aligned}\right\} \quad (1.55)$$

Eliminating λ_1 and λ_2 from equations (1.55), we find

$$x_1 - 2x_2 + x_3 = 0, \quad (1.56)$$

whence, using the constraint equation (1.49), we find $x_2 = 5/3$. Then it is easily deduced from equations (1.50) and (1.56) that either x_1 or x_3 is negative; thus the non-negativity restrictions are violated at this stationary point.

On the boundary $x_1 = 0$, equations (1.49) and (1.50) give

$$x_2, x_3 = \tfrac{1}{2}(5 \mp \sqrt{15}) = 0{\cdot}564,\ 4{\cdot}436,$$

where we have chosen $x_3 > x_2$ so as to make $f(\mathbf{x})$ as large as possible. We find

$$f(\mathbf{x}) = \tfrac{1}{2}(25 + \sqrt{15}) = 14{\cdot}436.$$

It is easily verified that $f(\mathbf{x}) < 14{\cdot}436$ on the remaining boundaries defined by the non-negativity restrictions. The final result is therefore

$$x_1^* = 0,\ x_2^* = \tfrac{1}{2}(5 - \sqrt{15}),\ x_3^* = \tfrac{1}{2}(5 + \sqrt{15});\ f(\mathbf{x}^*) = \tfrac{1}{2}(25 + \sqrt{15}).$$

1.5 THE LAGRANGIAN FUNCTION AND DUALITY

Consider again the classical optimization problem (1.8) and, in particular, the symmetrical form of the necessary conditions for a local maximum, equations (1.17). These equations show that the Lagrangian function $F(\mathbf{x},\lambda)$ is stationary with respect to both $\mathbf{x}$ and λ at the optimal point $(\mathbf{x}^*,\lambda^*)$, and this suggests that in solving the optimization problem (1.8) in the variables x_j we may also be solving a closely related optimization problem in the variables λ_i. We shall show that this is indeed the case, and that the function $F(\mathbf{x},\lambda)$ has a constrained local *minimum* with respect to λ at $(\mathbf{x}^*,\lambda^*)$, the constraints being equations (1.16). Furthermore, the value of this minimum is equal to the constrained local maximum value of $f(\mathbf{x})$ in problem (1.8).

Following Courant and Hilbert,[20] we assume that

$$\left.\begin{array}{l}\text{(a) } f(\mathbf{x}) \text{ has an actual constrained local maximum } z^* = f(\mathbf{x}^*), \text{ and} \\ \text{(b) if } \lambda \text{ is chosen arbitrarily in a neighbourhood of } \lambda^*, \text{ then} \\ \quad F(\mathbf{x},\lambda) \text{ has an actual local maximum } F(\mathbf{x}_0,\lambda) \text{ with respect to } \mathbf{x}.\end{array}\right\} \tag{1.57}$$

Assumption (b) implies that $\mathbf{x}_0$ satisfies the equations

$$\frac{\partial F}{\partial x_j} \equiv \frac{\partial f}{\partial x_j} - \sum_i \lambda_i \frac{\partial g_i}{\partial x_j} = 0; \tag{1.58}$$

we assume further that these equations determine $\mathbf{x}_0$ uniquely.

We can now prove the following important theorem.

***Theorem* 1.1**

In the classical optimization problem (1.8), *with the assumptions* (1.57), *the Lagrangian function* $F(\mathbf{x},\lambda)$ *has a saddle-point at the optimal point* $(\mathbf{x}^*,\lambda^*)$.

Proof

The Lagrangian function is

$$F(\mathbf{x},\lambda) \equiv f(\mathbf{x}) + \sum_i \lambda_i[b_i - g_i(\mathbf{x})].$$

Consider a neighbourhood of $(\mathbf{x}^*,\lambda^*)$ in E^{n+m}. Using assumption (1.57b) at $\lambda = \lambda^*$ (so that $\mathbf{x}_0 = \mathbf{x}^*$), we have

$$F(\mathbf{x},\lambda^*) \leqslant F(\mathbf{x}^*,\lambda^*) = f(\mathbf{x}^*) = F(\mathbf{x}^*,\lambda), \tag{1.59}$$

where the equalities are due to the fact that the constraints

$$g_i(\mathbf{x}) = b_i \tag{1.60}$$

are satisfied when $\mathbf{x} = \mathbf{x}^*$.

The relations (1.59) show that $F(\mathbf{x},\lambda)$ has a degenerate form of saddle-point at $(\mathbf{x}^*,\lambda^*)$; for an ordinary saddle-point at $(\mathbf{x}^*,\lambda^*)$ we would have

$$F(\mathbf{x},\lambda^*) \leqslant F(\mathbf{x}^*,\lambda^*) \leqslant F(\mathbf{x}^*,\lambda). \tag{1.61}$$

Theorem 1.1 leads directly to the idea of duality in classical optimization—the fundamental results are set out in Theorem 1.2.

***Theorem* 1.2**

Under the assumptions (1.57), *the problems*: (*i*) *maximize* $f(\mathbf{x})$, *subject to the constraints* (1.60) *and* (*ii*) *minimize* $F(\mathbf{x},\lambda)$, *subject to the constraints* (1.58), *may be regarded as dual problems. They have the property that in a neighbourhood of* $(\mathbf{x}^*,\lambda^*)$,

$$\max f(\mathbf{x}) = f(\mathbf{x}^*) = F(\mathbf{x}^*,\lambda^*) = \min F(\mathbf{x},\lambda), \tag{1.62}$$

where $f(\mathbf{x})$ *and* $F(\mathbf{x},\lambda)$ *are subject to their respective constraints* (1.60) *and* (1.58).

Proof

Define

$$h(\lambda) \equiv \max_{\mathbf{x}} F(\mathbf{x},\lambda) = F(\mathbf{x}_0,\lambda). \tag{1.63}$$

Assumption (1.57b) guarantees the existence of $h(\lambda)$ when λ is in a neighbourhood of λ^*; also, if $\lambda = \lambda^*$, then $\mathbf{x}_0 = \mathbf{x}^*$. Hence

$$h(\lambda^*) = F(\mathbf{x}^*,\lambda^*) = f(\mathbf{x}^*). \tag{1.64}$$

We shall now show that $h(\lambda)$ has a local minimum at $\lambda = \lambda^*$. Consider $F(\mathbf{x},\lambda)$ for a fixed value of λ. If $\mathbf{x}$ satisfies the constraints (1.60), then

$$F(\mathbf{x},\lambda) \equiv f(\mathbf{x})$$

and

$$\max_{\mathbf{x}} F(\mathbf{x},\lambda) = \max f(\mathbf{x}) = f(\mathbf{x}^*) = h(\lambda^*), \tag{1.65}$$

using equations (1.64). Equations (1.65) show that $h(\lambda) = h(\lambda^*)$ when $\mathbf{x}$ satisfies the constraints (1.60). Hence

$$h(\lambda) \geqslant h(\lambda^*) \tag{1.66}$$

when these constraints are removed, because the effect of removing the constraints is to enlarge the domain of $\mathbf{x}$ in a maximizing problem. Thus $h(\lambda)$ has a local minimum at $\lambda = \lambda^*$, i.e. $F(\mathbf{x},\lambda)$ has a local minimum with respect to λ at $\lambda = \lambda^*$, subject to the constraints (1.58) which determine $\mathbf{x}$ as a function of λ. However, for any fixed λ, the constraints (1.58) are necessary conditions for a local *maximum* of $F(\mathbf{x},\lambda)$ with respect to $\mathbf{x}$, and, in particular, for a local maximum of $f(\mathbf{x})$ when $\mathbf{x}$ satisfies the constraints (1.60). This completes the proof of equations (1.62).

It should be noted that in the derivation of equations (1.62) the minimization of $F(\mathbf{x},\lambda)$ is with respect to λ, where $\mathbf{x}$ is determined as a function of λ by the constraints (1.58); but these constraints may be used, in principle, to eliminate any set of n variables from the $(n+m)$ variables x_j and λ_i. The minimization of $F(\mathbf{x},\lambda)$ may therefore be regarded as being over *all* the variables x_j, λ_i, subject to the constraints (1.58).

Finally, from (1.63) and (1.66), we have

$$\max_{\mathbf{x}} F(\mathbf{x},\lambda) = h(\lambda), \quad \min h(\lambda) = h(\lambda^*). \tag{1.67}$$

It follows from (1.64) and (1.67) that

$$f(\mathbf{x}^*) = h(\lambda^*) = \min_{\lambda} \max_{\mathbf{x}} F(\mathbf{x},\lambda),$$

again showing the saddle-point property of $F(\mathbf{x},\lambda)$, though in a more general form than that of (1.59).

1.6 CONVEX AND CONCAVE FUNCTIONS

In general, we wish to find the global optimum of the objective function. A fundamental difficulty in the theory and practice of optimization is that most optimization techniques find only local optima—dynamic programming is a notable exception to this general statement. An obvious way out of this difficulty is to find every local optimum, and thence, by comparison, the global optimum; but this procedure is often time-consuming and sometimes impracticable. There is, however, a special class of functions, namely convex and concave functions, for which the local and global optima are closely related; we shall study the properties of these functions in the present section, beginning with some definitions.

The function $f(\mathbf{x})$ is said to be *convex* over a convex set X in E^n if, for any two points $\mathbf{x}_1, \mathbf{x}_2 \in X$ and for all $\lambda \in [0,1]$,

$$f[\lambda \mathbf{x}_2 + (1-\lambda)\mathbf{x}_1] \leqslant \lambda f(\mathbf{x}_2) + (1-\lambda) f(\mathbf{x}_1). \tag{1.68}$$

As a special case, the function $f(x)$ of the scalar x is convex in a domain X of x if (see Figure 1.1) PN$\leqslant$QN for all triads A,N,B. The function $f(x)$ of Figure 1.1 is often described as being *convex downwards*. Notice that the domain X of x must be a convex set; this ensures that if x_1 and x_2 belong

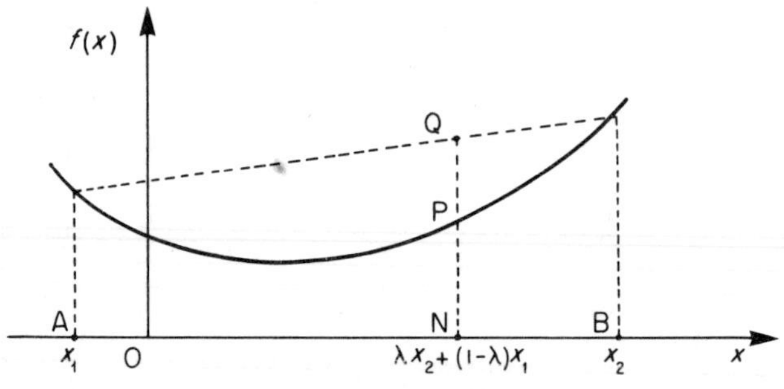

Figure 1.1 Convex function

to X then $\lambda x_2+(1-\lambda)x_1$ also belongs to X. More generally, the domain X of $\mathbf{x}$ must be a convex set.

The function $f(\mathbf{x})$ is said to be *concave* over a convex set X in E^n if, for any two points $\mathbf{x}_1,\mathbf{x}_2 \in X$ and for all $\lambda \in [0,1]$,

$$f[\lambda \mathbf{x}_2+(1-\lambda)\mathbf{x}_1] \geqslant \lambda f(\mathbf{x}_2)+(1-\lambda)f(\mathbf{x}_1). \tag{1.69}$$

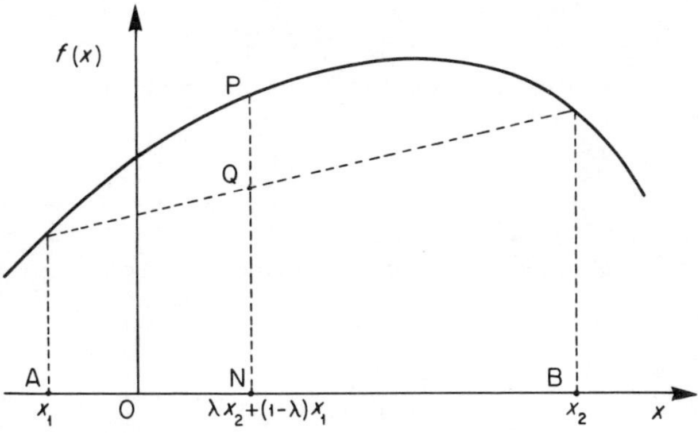

Figure 1.2 Concave function

Figure 1.2 shows the concave function $f(x)$ of the scalar x; geometrically, PN$\geqslant$QN for all triads A,N,B. Such a function is often described as being *concave downwards.*

If the inequalities in (1.68) and (1.69) are replaced by strict inequalities, then $f(\mathbf{x})$ is said to be *strictly convex* or *strictly concave*, respectively. For brevity, we shall usually omit any reference to the underlying convex set X unless we require X to have some specific property.

Note that in the definitions of a convex and a concave function we have not assumed that the function is continuous. It can be shown, however, that if a function is convex or concave over a convex set X and is bounded in X, then it is continuous at every interior point of X. A proof of this result for a function of one variable is given by Courant.[19]

The following elementary results are of considerable practical importance.

(a) If $f(\mathbf{x})$ is convex then $-f(\mathbf{x})$ is concave, and vice versa.

(b) The linear function $z=\mathbf{c}'\mathbf{x}$ is both convex and concave throughout E^n.

(c) A convex (concave) function has the property that its value at an interpolated point is less than (greater than) or equal to the value that would be obtained by linear interpolation.

(d) The sum of a finite number of convex (concave) functions is itself a convex (concave) function.

Proof for convex functions

Suppose the functions $f_k(\mathbf{x})$, $k=1,\ldots,N$, are convex, and let

$$f(\mathbf{x}) - \sum_{k=1}^{N} f_k(\mathbf{x}).$$

Then

$$\begin{aligned} f[\lambda\mathbf{x}_2 + (1-\lambda)\mathbf{x}_1] &= \sum_{k=1}^{N} f_k[\lambda\mathbf{x}_2 + (1-\lambda)\mathbf{x}_1] \\ &\leqslant \sum_{k=1}^{N} [\lambda f_k(\mathbf{x}_2) + (1-\lambda) f_k(\mathbf{x}_1)] \\ &= \lambda f(\mathbf{x}_2) + (1-\lambda) f(\mathbf{x}_1), \end{aligned}$$

as required. The proof for concave functions is similar.

Corollary

The proof shows that the result remains valid as $N \to \infty$, provided that the infinite series involved are absolutely convergent.

(e) The positive semidefinite quadratic form $z = \mathbf{x}'\mathbf{A}\mathbf{x}$ is a convex function throughout E^n.

Proof

Without loss of generality we may assume that $\mathbf{A}$ is symmetric. Let $\hat{\mathbf{x}} = \lambda\mathbf{x}_2 + (1-\lambda)\mathbf{x}_1$ for any vectors $\mathbf{x}_1, \mathbf{x}_2 \in E^n$. Then

$$\begin{aligned} \hat{\mathbf{x}}'\mathbf{A}\hat{\mathbf{x}} &= [\lambda\mathbf{x}_2 + (1-\lambda)\mathbf{x}_1]'\mathbf{A}[\lambda\mathbf{x}_2 + (1-\lambda)\mathbf{x}_1] \\ &= [\mathbf{x}_1 + \lambda(\mathbf{x}_2 - \mathbf{x}_1)]'\mathbf{A}[\mathbf{x}_1 + \lambda(\mathbf{x}_2 - \mathbf{x}_1)] \\ &= \mathbf{x}_1'\mathbf{A}\mathbf{x}_1 + 2\lambda(\mathbf{x}_2 - \mathbf{x}_1)'\mathbf{A}\mathbf{x}_1 + \lambda^2(\mathbf{x}_2 - \mathbf{x}_1)'\mathbf{A}(\mathbf{x}_2 - \mathbf{x}_1). \end{aligned}$$

But $\mathbf{x}'\mathbf{A}\mathbf{x} \geqslant 0$ for all $\mathbf{x}$; also $\lambda \in [0,1]$. Therefore

$$\lambda^2\mathbf{x}'\mathbf{A}\mathbf{x} \leqslant \lambda\mathbf{x}'\mathbf{A}\mathbf{x} \qquad \text{for all } \mathbf{x}, \tag{1.70}$$

and hence

$$\begin{aligned} \hat{\mathbf{x}}'\mathbf{A}\hat{\mathbf{x}} &\leqslant \mathbf{x}_1'\mathbf{A}\mathbf{x}_1 + 2\lambda(\mathbf{x}_2 - \mathbf{x}_1)'\mathbf{A}\mathbf{x}_1 + \lambda(\mathbf{x}_2 - \mathbf{x}_1)'\mathbf{A}(\mathbf{x}_2 - \mathbf{x}_1) \\ &= \mathbf{x}_1'\mathbf{A}\mathbf{x}_1 + \lambda(\mathbf{x}_2 - \mathbf{x}_1)'\mathbf{A}\mathbf{x}_1 + \lambda(\mathbf{x}_2 - \mathbf{x}_1)'\mathbf{A}\mathbf{x}_2 \\ &= \lambda\mathbf{x}_2'\mathbf{A}\mathbf{x}_2 + (1-\lambda)\mathbf{x}_1'\mathbf{A}\mathbf{x}_1, \end{aligned}$$

as required.

(f) A negative semidefinite quadratic form is a concave function throughout E^n. This follows from (a) and (e).

(g) The positive definite quadratic form $z = \mathbf{x}'\mathbf{A}\mathbf{x}$ is a strictly convex function throughout E^n, for in this case $\mathbf{x}'\mathbf{A}\mathbf{x} > 0$ for all $\mathbf{x} \neq \mathbf{0}$. Hence, for $\lambda \in (0,1)$, the strict inequality

$$\lambda^2\mathbf{x}'\mathbf{A}\mathbf{x} < \lambda\mathbf{x}'\mathbf{A}\mathbf{x} \quad \text{for all } \mathbf{x} \neq \mathbf{0}$$

replaces (1.70); the rest of the proof follows as in (e) above.

(h) A negative definite quadratic form is a strictly concave function throughout E^n. This follows from (a) and (g).

Further important properties of convex and concave functions are proved in Theorems 1.3 to 1.8.

***Theorem* 1.3**

Let $f(\mathbf{x}) \in C_1$ throughout the interior of the convex set X in E^n, and suppose that $f(\mathbf{x})$ is convex over X. Then

$$(\mathbf{x}_2 - \mathbf{x}_1)'\nabla f(\mathbf{x}_1) \leqslant f(\mathbf{x}_2) - f(\mathbf{x}_1) \tag{1.71}$$

for all interior points $\mathbf{x}_1 \in X$ and all points $\mathbf{x}_2 \in X$.

Proof

Since $f(\mathbf{x})$ is convex over X, we have

$$f[\lambda\mathbf{x}_2 + (1-\lambda)\mathbf{x}_1] \leqslant \lambda f(\mathbf{x}_2) + (1-\lambda) f(\mathbf{x}_1)$$

for all $\lambda \in [0, 1]$. Therefore

$$\frac{f[\mathbf{x}_1 + \lambda(\mathbf{x}_2 - \mathbf{x}_1)] - f(\mathbf{x}_1)}{\lambda} \leqslant f(\mathbf{x}_2) - f(\mathbf{x}_1). \tag{1.72}$$

By the first mean value theorem,

$$f[\mathbf{x}_1 + \lambda(\mathbf{x}_2 - \mathbf{x}_1)] = f(\mathbf{x}_1) + \lambda(\mathbf{x}_2 - \mathbf{x}_1)'\nabla f[\mathbf{x}_1 + \theta\lambda(\mathbf{x}_2 - \mathbf{x}_1)]$$

for some $\theta \in [0, 1]$, so that (1.72) becomes

$$(\mathbf{x}_2 - \mathbf{x}_1)'\nabla f[\mathbf{x}_1 + \theta\lambda(\mathbf{x}_2 - \mathbf{x}_1)] \leqslant f(\mathbf{x}_2) - f(\mathbf{x}_1). \tag{1.73}$$

The required inequality (1.71) follows from (1.73) as $\lambda \to 0$.

Note. The inequality (1.71) is obvious geometrically when f is a function of a single scalar variable.

***Theorem* 1.4**

Suppose that $f(\mathbf{x})$ is a convex function for all $\mathbf{x} \geqslant 0$, and let V be the non-empty set

$$V = \{\mathbf{x} : f(\mathbf{x}) \leqslant b,\ \mathbf{x} \geqslant 0\}.$$

Then V is a convex set.

Proof

The case where V contains only one point is trivial. Otherwise, let $\mathbf{x}_1, \mathbf{x}_2 \in V$. Then

$$\lambda\mathbf{x}_2 + (1-\lambda)\mathbf{x}_1 \geqslant 0 \quad \text{if} \quad \lambda \in [0, 1],$$

and

$$\begin{aligned} f[\lambda\mathbf{x}_2 + (1-\lambda)\mathbf{x}_1] &\leqslant \lambda f(\mathbf{x}_2) + (1-\lambda) f(\mathbf{x}_1) \\ &\leqslant \lambda b + (1-\lambda) b \\ &= b. \end{aligned}$$

Thus

$$\lambda \mathbf{x}_2 + (1-\lambda)\mathbf{x}_1 \subset V \quad \text{if} \quad \mathbf{x}_1, \mathbf{x}_2 \in V,$$

i.e. V is convex.

Note. The set of points $W = \{\mathbf{x} : f(\mathbf{x}) = b, \mathbf{x} \geqslant \mathbf{0}\}$ is *not*, in general, a convex set.

Before stating the next theorem we need some definitions. A *hyperplane* is a set of points

$$H = \{\mathbf{x} : \mathbf{a}'\mathbf{x} = c\},$$

where $\mathbf{a}$ ($\neq \mathbf{0}$) and c are constants. Now let $\mathbf{x}_b$ be a boundary point of a convex set X. Then

$$\mathbf{a}'\mathbf{x} = c$$

is called a *supporting hyperplane* to X at $\mathbf{x}_b$ if $\mathbf{a}'\mathbf{x}_b = c$ and if all of X lies in one closed half-space defined by the hyperplane, i.e. if

$$\mathbf{a}'\mathbf{x} \leqslant c \text{ for all } \mathbf{x} \in X \quad \text{or} \quad \mathbf{a}'\mathbf{x} \geqslant c \text{ for all } \mathbf{x} \in X.$$

***Theorem* 1.5**

Suppose that $f(\mathbf{x})$ is a convex function for all $\mathbf{x} \geqslant \mathbf{0}$ and that $f(\mathbf{x}) \in C_1$. Then the tangent hyperplane to the hypersurface $f(\mathbf{x}) = b$ at the point $\mathbf{x}_0 \geqslant \mathbf{0}$ is a supporting hyperplane at $\mathbf{x}_0$ to the convex set

$$K = \{\mathbf{x} : f(\mathbf{x}) \leqslant b, \mathbf{x} \geqslant \mathbf{0}\}.$$

Proof

The tangent hyperplane to $f(\mathbf{x}) = b$ at $\mathbf{x}_0$ is

$$[\nabla f(\mathbf{x}_0)]'\mathbf{x} = [\nabla f(\mathbf{x}_0)]'\mathbf{x}_0.$$

It is therefore sufficient to prove that if $\mathbf{x}_2 \in K$, then

$$[\nabla f(\mathbf{x}_0)]'\mathbf{x}_2 \leqslant [\nabla f(\mathbf{x}_0)]'\mathbf{x}_0. \tag{1.74}$$

Setting $\mathbf{x}_1 = \mathbf{x}_0$ in the inequality (1.71), we obtain

$$(\mathbf{x}_2 - \mathbf{x}_0)'\nabla f(\mathbf{x}_0) \leqslant f(\mathbf{x}_2) - f(\mathbf{x}_0) \leqslant b - b = 0,$$

which is identical with (1.74).

Theorems 1.6 to 1.8 give results for concave functions analogous to those of Theorems 1.3 to 1.5 for convex functions. The proofs are similar.

***Theorem* 1.6**

Let $f(\mathbf{x}) \in C_1$ throughout the interior of the convex set X in E^n and suppose that $f(\mathbf{x})$ is concave over X. Then

$$(\mathbf{x}_2 - \mathbf{x}_1)'\nabla f(\mathbf{x}_1) \geqslant f(\mathbf{x}_2) - f(\mathbf{x}_1) \tag{1.75}$$

for all interior points $\mathbf{x}_1 \in X$ *and all points* $\mathbf{x}_2 \in X$.

***Theorem* 1.7**

Suppose that $f(\mathbf{x})$ *is a concave function for all* $\mathbf{x} \geqslant \mathbf{0}$, *and let* V *be the non-empty set*

$$V = \{\mathbf{x} : f(\mathbf{x}) \geqslant b, \ \mathbf{x} \geqslant \mathbf{0}\}.$$

Then V *is a convex set.*

***Theorem* 1.8**

Suppose that $f(\mathbf{x})$ *is a concave function for all* $\mathbf{x} \geqslant \mathbf{0}$ *and that* $f(\mathbf{x}) \in C_1$. *Then the tangent hyperplane to the hypersurface* $f(\mathbf{x}) = b$ *at the point* $\mathbf{x}_0 \geqslant \mathbf{0}$ *is a supporting hyperplane at* $\mathbf{x}_0$ *to the convex set*

$$K = \{\mathbf{x} : f(\mathbf{x}) \geqslant b, \ \mathbf{x} \geqslant \mathbf{0}\}.$$

Examples of convex and concave functions

(i) e^x is a convex function for all x.

(ii) $\sin x$ is neither a convex nor a concave function over $0 \leqslant x \leqslant 2\pi$. However, it is concave for $0 \leqslant x \leqslant \pi$ and convex for $\pi \leqslant x \leqslant 2\pi$.

(iii) The separable function $\sum_j f_j(x_j)$ is convex if each $f_j(x_j)$ is convex.

(iv) If $f(x) \in C_2$, the definitions (1.68) and (1.69) are equivalent to $f''(x) \geqslant 0$ and $f''(x) \leqslant 0$, respectively, at all points of an appropriate convex set X. The proofs are left to the exercises.

(v) If $f(\mathbf{x})$ and $g(\mathbf{x})$ are convex functions, then $f(\mathbf{x})g(\mathbf{x})$ is not necessarily a convex function. For example, the functions x^2 and $(x-a)^2$ are both convex, though $x^2(x-a)^2$ is neither convex nor concave.

The following four theorems, 1.9 to 1.12, establish the main results concerning the maxima and minima of convex functions; Theorems 1.13 to 1.16 provide similar results for concave functions.

***Theorem* 1.9**

Let $f(\mathbf{x})$ *be a convex function over the closed convex set* X *in* E^n. *Then any local minimum of* $f(\mathbf{x})$ *is also the global minimum of* $f(\mathbf{x})$ *over* X.

Proof

The proof is by contradiction. Assume that $f(\mathbf{x})$ takes on a local minimum at $\mathbf{x}_L \in X$, that its global minimum is at $\mathbf{x}^* \in X$, and that $f(\mathbf{x}^*) < f(\mathbf{x}_L)$. Now

$$\begin{aligned} f[\lambda\mathbf{x}^* + (1-\lambda)\mathbf{x}_L] &\leqslant \lambda f(\mathbf{x}^*) + (1-\lambda)f(\mathbf{x}_L) \\ &< \lambda f(\mathbf{x}_L) + (1-\lambda)f(\mathbf{x}_L) \\ &= f(\mathbf{x}_L) \end{aligned} \tag{1.76}$$

for all $\lambda \in (0, 1)$. But for sufficiently small λ, the point

$$\mathbf{x} = \lambda\mathbf{x}^* + (1-\lambda)\mathbf{x}_L$$

lies in the ε-neighbourhood of $\mathbf{x}_L$, and (1.76) then shows that $f(\mathbf{x}) < f(\mathbf{x}_L)$ in this neighbourhood, which contradicts the fact that $f(\mathbf{x})$ has a local minimum at $\mathbf{x} = \mathbf{x}_L$. Thus $\mathbf{x}_L$ and $\mathbf{x}^*$ cannot be distinct.

Theorem 1.9 is intuitively obvious. If the function $f(\mathbf{x})$ increases between the points $\mathbf{x}^*$ and $\mathbf{x}_L$ then, because it is convex, it must increase between some point in the ε-neighbourhood of $\mathbf{x}_L$ and the point $\mathbf{x}_L$ itself – but this is impossible by the definition of $\mathbf{x}_L$.

***Theorem* 1.10**

Let $f(\mathbf{x})$ be a convex function over the closed convex set X in E^n. Then the set of points at which $f(\mathbf{x})$ takes on its global minimum is a convex set.

Proof

The case where $f(\mathbf{x})$ takes on its global minimum at a single point is trivial. Otherwise, suppose the global minimum is taken on at $\mathbf{x}_1$ and $\mathbf{x}_2$ ($\neq \mathbf{x}_1$), and let

$$\mathbf{x} = \lambda \mathbf{x}_2 + (1-\lambda)\mathbf{x}_1, \qquad \lambda \in [0,1]. \tag{1.77}$$

Then

$$\begin{aligned} f(\mathbf{x}) &= f[\lambda \mathbf{x}_2 + (1-\lambda)\mathbf{x}_1] \\ &\leqslant \lambda f(\mathbf{x}_2) + (1-\lambda) f(\mathbf{x}_1). \end{aligned}$$

But $f(\mathbf{x}_2) = f(\mathbf{x}_1) = f(\mathbf{x}^*)$, the global minimum. Hence

$$f(\mathbf{x}) = f(\mathbf{x}^*)$$

for all points $\mathbf{x}$ defined by (1.77), which proves the theorem.

Theorem 1.10 has the following important corollaries.

***Corollary* 1**

If the global minimum of $f(\mathbf{x})$ is taken on at two distinct points, then it is taken on at every point in the line segment joining them—cf. alternative optima in linear programming.[75]

***Corollary* 2**

There cannot be two (or more) points at which $f(\mathbf{x})$ takes on a strong *local minimum.*

Proof

Any local minimum is also the global minimum, by Theorem 1.9. Then, by Corollary 1, if the global minimum is taken on at two distinct points, it cannot be a strong minimum.

***Corollary* 3**

If $f(\mathbf{x})$ is strictly *convex, then the global minimum of $f(\mathbf{x})$ is taken on at a unique point.*

***Corollary* 2**

There cannot be two (or more) points at which $f(\mathbf{x})$ takes on a strong *local maximum.*

***Corollary* 3**

If $f(\mathbf{x})$ is strictly *concave, then the global maximum of $f(\mathbf{x})$ is taken on at a unique point.*

***Theorem* 1.15**

Let $f(\mathbf{x}) \in C_1$ throughout the interior of the convex set X in E^n and suppose that $f(\mathbf{x})$ is concave over X. If $\nabla f(\mathbf{x}_0) = 0$, where $\mathbf{x}_0$ is an interior point of X, then $f(\mathbf{x})$ takes on its global maximum over X at $\mathbf{x} = \mathbf{x}_0$.

***Theorem* 1.16**

Let X be a closed convex set which is bounded from below and suppose that $f(\mathbf{x})$ is concave over X in E^n. If the global minimum *of $f(\mathbf{x})$ is finite, then it will be taken on at one or more of the extreme points of X.*

SUMMARY

A useful first step in tackling an optimization problem is to classify the problem and, in particular, to discover whether some specialized algorithm is appropriate for its solution. The main classifications are given in Section 1.2. In Section 1.3, the classical optimization problem (every constraint an equation) is solved by the standard Lagrange multiplier technique. In Section 1.4, this method is extended to the case where inequality constraints are present. Section 1.5 deals with the saddle-point property of the Lagrangian function (Theorem 1.1), which leads to the idea of duality in classical optimization; these topics may be regarded as an introduction to Kuhn–Tucker theory, which is developed in detail in Chapter 2. Finally, some important theorems on convex and concave functions are given in Section 1.6. One of the most useful of these, as well as the simplest, is Theorem 1.9, which states that any local minimum of a convex function is also its global minimum.

EXERCISES

1. Find the extreme values of each of the following functions (i) by eliminating one of the variables and (ii) by the Lagrange multiplier method. Determine in each case whether these values are maxima or minima.
   (a) $z = x_1^2 + x_2^2$, subject to $x_1 x_2 = 2$.
   (b) $z = x_1 x_2$, subject to $x_1^2 + x_2^2 = 2$.
2. Find the maximum and minimum values of

$$z = x_1^2 + x_2^2 + x_3^2,$$

subject to $x_1 + 2x_2 = 2$, $x_1^2 + 2x_2^2 + 2x_3^2 = 2$.

3. Find the maximum value of xy, given that the point $[x, y]$ lies on the Witch of Agnesi $xy^2 = a^2(a-x)$.

4. Prove by vector methods that the minimum distance from the origin to the line of intersection of the planes

$$a_1x + a_2y + a_3z = p_1,$$
$$b_1x + b_2y + b_3z = p_2$$

is

$$|p_1\mathbf{b} - p_2\mathbf{a}|/|\mathbf{a} \times \mathbf{b}|,$$

where $\mathbf{a} = [a_1, a_2, a_3]$, $\mathbf{b} = [b_1, b_2, b_3]$. Verify that the Lagrange multiplier method gives the same result.

5. Maximize $z = |\mathbf{a}'\mathbf{x}|$, subject to $|\mathbf{x}| = 1$, using the Lagrange multiplier method. Verify your answer by using Schwartz's inequality.

6. Maximize $z = x_1x_2....x_n$, subject to $x_1^2 + x_2^2 + ... + x_n^2 = 1$. Deduce that the geometric mean of n positive numbers is not greater than their arithmetic mean.

7. Investigate the maxima and minima of the function

$$z = \cos A \cos B \cos C,$$

given that $A + B + C = 0$.

8. Prove that the function

$$z = ax_2x_3 + bx_3x_1 + cx_1x_2,$$

subject to

$$x_1 + x_2 + x_3 = 1,$$

has a stationary value abc/d, where

$$d = 2(bc + ca + ab) - (a^2 + b^2 + c^2) \neq 0.$$

Prove also that if $d > 0$ this value is a maximum or minimum according as a, b, c are all positive or all negative, and that if $d < 0$ it is neither a maximum nor a minimum.

9. Let $\Delta = \det\{x_{ij}\}$, where the elements x_{ij} are all real, let $\mathbf{x}_j = [x_{1j}, ..., x_{nj}]$ be the jth column of Δ and let $|\mathbf{x}_j| = d_j$. Prove Hadamard's theorem:

$$\Delta^2 \leqslant d_1^2 d_2^2 .... d_n^2,$$

where the equality holds if and only if the matrix $\{x_{ij}\}$ is orthogonal.

10. Find the shortest distance from the point (r_1, θ_1) to the lemniscate $r^2 = 2a^2 \cos 2\theta$.

11. Without loss of generality, the matrix $\mathbf{A}$ in equation (1.28) may be assumed symmetric. Prove that every root of this equation is real. [The proof is similar to the one which shows that every latent root of a real symmetric matrix is real.]

12. Minimize $z = x_1^2 + 4x_2^2 - 2x_1 + 8x_2$, subject to

$$5x_1 + 2x_2 \leqslant 4,$$
$$x_1, x_2 \geqslant 0.$$

13. Maximize $z = x_1 x_2 x_3$, subject to

$$10x_1^2 + x_2^2 + 4x_3^2 \leqslant 40,$$
$$x_1^2 + x_2^2 + x_3^2 \leqslant 15,$$
$$x_1, x_2, x_3 \geqslant 0.$$

14. Discuss the problem of maximizing the function

$$z = x - \frac{2}{x} - x^2,$$

subject to $x \geqslant k$, where $k(\neq 0)$ is a constant.

15. Discuss the convexity and concavity of the following functions:

(a) $x_1^2 + 3x_2^2 + 9x_3^2 - 2x_1x_2 + 6x_2x_3 + 2x_3x_1$.

(b) $|\mathbf{x}|$.

(c) $x_1 f_1(x_1) + x_2 f_2(x_2)$, where $x_1, x_2 \geqslant 0$ and both $f_1(x_1)$ and $f_2(x_2)$ are convex functions.

(d) $\int_a^b f_1(\mathbf{x}) g_1(y) dy + \int_c^d f_2(\mathbf{x}) g_2(y) dy$, where $g_1(y) \geqslant 0$, $g_2(y) \geqslant 0$ and both $f_1(\mathbf{x})$ and $f_2(\mathbf{x})$ are concave functions.

(e) $\int_0^1 f(\mathbf{x}, y) g(y) dy$, where $g(y) \geqslant 0$ and $f(\mathbf{x}, y)$ is a convex function of $\mathbf{x}$ for $0 \leqslant y \leqslant 1$.

16. If $h \geqslant k \geqslant 0$ and if $f(x)$ is convex for $x_0 - h \leqslant x \leqslant x_0 + h$, show that

$$f(x_0 + k) + f(x_0 - k) \leqslant f(x_0 + h) + f(x_0 - h).$$

17. If $f(x)$ is convex for $x_0 \leqslant x \leqslant x_0 + (n+1)h$, show that for $n = 1, 2, \ldots.,$

$$\frac{f(x_0 + nh) - f(x_0)}{n} \leqslant \frac{f[x_0 + (n+1)h] - f(x_0)}{n+1}.$$

18. Let $f(\mathbf{x})$ be a convex function for all $\mathbf{x} \geqslant \mathbf{0}$. Give examples to show that the set of points

$$W = \{\mathbf{x} : f(\mathbf{x}) = b, \ \mathbf{x} \geqslant \mathbf{0}\}$$

may or may not be a convex set.

19. Give an example, if one exists, of each of the following types of function. (Unless otherwise stated, the underlying convex set X is open to choice.)

(a) Convex and differentiable throughout E^n.

(b) Convex throughout E^n and not differentiable at $\mathbf{x} = \mathbf{0}$.

(c) Concave over X and discontinuous at $\mathbf{x} = \mathbf{x}_0$, where $\mathbf{x}_0 \in X$.

(d) Convex, concave and nonlinear over X.

20. If $f(x) \in C_2$, prove that the expression

$$\lambda f(x_2) + (1-\lambda) f(x_1) - f[\lambda x_2 + (1-\lambda) x_1]$$

can be written as the double integral

$$\lambda (x_2 - x_1)^2 \int_0^1 \mathrm{d}t \int_{\lambda t}^t f''[x_1 + (x_2 - x_1)u] \mathrm{d}u.$$

Deduce that $f(x)$ is convex or concave if and only if $f''(x) \geqslant 0$ or $f''(x) \leqslant 0$, respectively, at all points of an appropriate convex set X.

21. Prove that if the quadratic form $z = \mathbf{x}'\mathbf{A}\mathbf{x}$ is convex throughout E^n, then $\mathbf{A}$ is positive semidefinite. What are the corresponding results for the cases in which z is (a) concave, (b) strictly convex, (c) strictly concave?

22. If $f(\mathbf{x}) \equiv \frac{1}{2}\mathbf{x}'\mathbf{A}\mathbf{x} + \mathbf{b}'\mathbf{x}$ is strictly convex over a convex set X, where X is a subset of E^n, what can be said about the matrix $\mathbf{A}$?

23. The range $R(\mathbf{A})$ of a matrix $\mathbf{A}$ is defined as the vector space spanned by its columns and the null space $N(\mathbf{A})$ of $\mathbf{A}$ is defined as the vector space spanned by the vectors orthogonal to its columns. Prove that:

(a) If $\mathbf{y} \in R(\mathbf{A})$, then there exists an $\mathbf{x}$ such that $\mathbf{y} = \mathbf{A}\mathbf{x}$.

(b) If $\mathbf{w} \in N(\mathbf{A})$, then $\mathbf{A}'\mathbf{w} = \mathbf{0}$.

(c) If $\mathbf{y} \in R(\mathbf{A})$ and $\mathbf{w} \in N(\mathbf{A})$, then $\mathbf{y}'\mathbf{w} = 0$.

Now consider the unconstrained quadratic function

$$f(\mathbf{x}) \equiv \tfrac{1}{2}\mathbf{x}'\mathbf{A}\mathbf{x} + \mathbf{b}'\mathbf{x}.$$

Prove that:

(d) If $\mathbf{A}$ is negative definite, then $f(\mathbf{x})$ has a strong global maximum.

(e) If $\mathbf{A}$ is negative semidefinite and if $\mathbf{b} \in R(\mathbf{A})$, then $f(\mathbf{x})$ has a weak global maximum.

(f) If $f(\mathbf{x})$ has a weak global maximum at $\mathbf{x} = \mathbf{x}^*$ and if $\mathbf{w} \in N(\mathbf{A})$, then $f(\mathbf{x})$ also has a weak global maximum at $\mathbf{x} = \mathbf{x}^* + \mathbf{w}$.

CHAPTER 2

Nonlinear Programming

2.1 INTRODUCTION

In 1951, Kuhn and Tucker[51] produced some results which gave a new insight into the nature of the optimal solutions of nonlinear programming problems. For the general nonlinear programming problem, they obtained necessary conditions for a local optimum and, in certain important special cases, necessary and sufficient conditions for a global optimum. Their conditions are related to the saddle-point property of the Lagrangian function at the optimal point (Theorem 1.1, page 20).

In the next three sections, we shall derive the main results of Kuhn–Tucker theory. In Sections 2.5 to 2.7, we consider the quadratic programming problem; in many cases of practical interest it turns out that the application of Kuhn–Tucker theory to this problem leads to a very efficient algorithm for its solution. Finally, in Section 2.8, we describe Griffith and Stewart's method for solving nonlinear programming problems; this method involves the solution of a sequence of linear programming problems. An excellent survey of nonlinear programming methods up to 1963 is given by Wolfe.[79]

2.2 KUHN–TUCKER NECESSARY CONDITIONS

Consider the general nonlinear programming problem:

maximize

$$z = f(\mathbf{x}),$$

subject to

$$\left.\begin{aligned} g_i(\mathbf{x}) &\leqslant b_i \quad (i=1,\ldots..u),\\ g_i(\mathbf{x}) &\geqslant b_i \quad (i=u+1,\ldots..v),\\ g_i(\mathbf{x}) &= b_i \quad (i=v+1,\ldots.,m),\\ \mathbf{x} &\geqslant 0. \end{aligned}\right\} \tag{2.1}$$

Any vector $\mathbf{x}$ satisfying the constraints and non-negativity restrictions in problem (2.1) will be called a *feasible solution* for the problem. We shall assume throughout this chapter that any minimizing problem has been reformulated as a maximizing problem. This will avoid duplication of theorems in the text; the reader is invited to write out for himself the results for minimizing problems.

If $f(\mathbf{x})$ in problem (2.1) takes on a local maximum, subject to the constraints and non-negativity restrictions, at the point $\mathbf{x} = \mathbf{x}^*$, then the Kuhn–Tucker necessary conditions are conditions which $\mathbf{x}^*$ must satisfy. We shall now derive these conditions.

Assume that f, $g_i \in C_1$, and define the index sets

$$\begin{aligned} I_a &= \{i : g_i(\mathbf{x}^*) = b_i\}, \\ I_p &= \{i : g_i(\mathbf{x}^*) \neq b_i\}, \\ J_a &= \{j : x_j^* = 0\}, \\ J_p &= \{j : x_j^* > 0\}. \end{aligned}$$

The suffix a on I and J refers to *active* constraints and non-negativity restrictions; the suffix p refers to *passive* constraints and non-negativity restrictions.

In problem (2.1), regard the non-negativity restrictions $\mathbf{x} \geqslant \mathbf{0}$ as a set of n explicit constraints, making $(m+n)$ constraints in all. Define the *extended Lagrangian function*

$$F_e(\mathbf{x}, \boldsymbol{\lambda}) \equiv f(\mathbf{x}) + \sum_i \lambda_i [b_i - g_i(\mathbf{x})] - \sum_j \lambda_{m+j} x_j, \tag{2.2}$$

where $\lambda_1, \ldots, \lambda_{m+n}$ are Lagrange multipliers (sometimes called *generalized Lagrange multipliers*, or *Kuhn–Tucker multipliers*) associated with these $(m+n)$ constraints—cf. the Lagrangian function F of (1.12). The following theorem leads directly to the Kuhn–Tucker necessary conditions.

***Theorem* 2.1**

For problem (2.1) *the m Lagrange multipliers* λ_i *in the extended Lagrangian function* (2.2) *satisfy the following conditions*:

$$\frac{\partial f(\mathbf{x}^*)}{\partial x_j} - \sum_i \lambda_i^* \frac{\partial g_i(\mathbf{x}^*)}{\partial x_j} \leqslant 0 \quad (j \in J_a) \tag{2.3}$$

$$= 0 \quad (j \in J_p); \tag{2.4}$$

$$\left.\begin{aligned} \lambda_i^* &\geqslant 0 \quad (i = 1, \ldots, u) \\ &\leqslant 0 \quad (i = u+1, \ldots, v) \\ &\text{is unrestricted in sign } (i = v+1, \ldots, m). \end{aligned}\right\} \tag{2.5}$$

Proof

A necessary condition for a constrained local maximum of $f(\mathbf{x})$ in problem (2.1) is that there exist $(m+n)$ Lagrange multipliers λ_i^*, λ_{m+j}^* satisfying the equations

$$\frac{\partial F_e}{\partial x_j}(\mathbf{x}^*, \boldsymbol{\lambda}^*) \equiv \frac{\partial f(\mathbf{x}^*)}{\partial x_j} - \sum_i \lambda_i^* \frac{\partial g_i(\mathbf{x}^*)}{\partial x_j} - \lambda_{m+j}^* = 0. \tag{2.6}$$

This follows from the theory of classical optimization (Section 1.3) after slack and surplus variables have been added to the constraints and non-negativity restrictions of problem (2.1). Note, however, that the slack and surplus variables do not appear in equations (2.6).

We have already seen (page 17) that passive constraints are associated with Lagrange multipliers that take the value zero at the optimal point. In particular, a passive non-negativity restriction $x_j^* > 0$ is associated with $\lambda_{m+j}^* = 0\ (j \in J_p)$. Thus equations (2.4) follow immediately from equations (2.6).

To prove the inequalities (2.3), we rewrite the non-negativity restrictions $x_j \geqslant 0$ as $x_j \geqslant b_{m+j}$, where $b_{m+j} = 0$, and then suppose that each b_{m+j} is reduced from zero to a small negative number. This enlarges the feasible region, and so z^* can only increase or remain unchanged, i.e.

$$\frac{\partial z^*}{\partial b_{m+j}} \leqslant 0.$$

Equation (1.46) then shows that $\lambda_{m+j}^* \leqslant 0$, and the inequalities (2.3) follow from equations (2.6).

By considering similar variations in the b_i, it can be shown that the λ_i^* satisfy conditions (2.5). The details are left as an exercise.

From the proof of Theorem 2.1, we see that the signs of the λ_i^* and λ_{m+j}^* are easily remembered: they are such that each of the corresponding terms in the extended Lagrangian function $F_e(\mathbf{x}^*, \boldsymbol{\lambda}^*)$ is non-negative.

The proof of Theorem 2.1 is invalid if the Lagrange multipliers λ_i^*, λ_{m+j}^* do not exist. Two comments are relevant here. First, if the λ_i^* exist when the non-negativity restrictions are absent, then the λ_i^* *and* λ_{m+j}^* exist when they are present, for the necessary and sufficient existence condition in the absence of non-negativity restrictions,

$$\operatorname{rank}(\mathbf{G}) = \operatorname{rank}(\mathbf{G} \,\vdots\, \nabla f), \tag{2.7}$$

becomes

$$\operatorname{rank}(\mathbf{G} \,\vdots\, \mathbf{I}_n) = \operatorname{rank}(\mathbf{G} \,\vdots\, \mathbf{I}_n \,\vdots\, \nabla f) \tag{2.8}$$

when they are present, and the latter condition is obviously implied by the former. (The $(n \times m)$ matrix $\mathbf{G}$ in (2.7) and (2.8) is defined by equation (1.39) and $\mathbf{I}_n$ is the unit matrix of order n). Secondly, the question of the existence of the Lagrange multipliers λ_i^* is considered in more detail in Section 2.4, where the Kuhn–Tucker constraint qualification is discussed.

With the help of Theorem 2.1, we can write down the Kuhn–Tucker necessary conditions for problem (2.1). We define the Lagrangian function

$$F(\mathbf{x}, \boldsymbol{\lambda}) \equiv f(\mathbf{x}) + \sum_i \lambda_i [b_i - g_i(\mathbf{x})]. \tag{2.9}$$

Then, if $f(\mathbf{x})$ takes on a constrained local maximum at the point $\mathbf{x} = \mathbf{x}^*$, it is necessary that a vector $\boldsymbol{\lambda}^*$ exists such that:

Kuhn–Tucker condition I (K–T I)

$$\nabla_x F(\mathbf{x}^*,\boldsymbol{\lambda}^*) \equiv \nabla f(\mathbf{x}^*) - \sum_i \lambda_i{}^* \nabla g_i(\mathbf{x}^*) \leqslant \mathbf{0}, \tag{2.10}$$

where the strict equality holds for components $j \in J_p$. This follows from (2.3) and (2.4). Also,

Kuhn–Tucker condition II (K–T II)

$$[\nabla_x F(\mathbf{x}^*,\boldsymbol{\lambda}^*)]'\mathbf{x}^* \equiv \sum_j \left\{\frac{\partial f}{\partial x_j}(\mathbf{x}^*) - \sum_i \lambda_i{}^* \frac{\partial g_i}{\partial x_j}(\mathbf{x}^*)\right\} x_j{}^* = 0, \tag{2.11}$$

since either

$$\frac{\partial f}{\partial x_j}(\mathbf{x}^*) - \sum_i \lambda_i{}^* \frac{\partial g_i}{\partial x_j}(\mathbf{x}^*) = 0 \quad (j \in J_p),$$

from (2.4), or

$$x_j{}^* = 0 \quad (j \in J_a),$$

from the definition of J_a.

The third Kuhn–Tucker necessary condition is merely a restatement of the constraints in problem (2.1). Using the Lagrangian function (2.9), we obtain

Kuhn–Tucker condition III (K–T III)

The first u components of the vector

$$\nabla_\lambda F(\mathbf{x}^*,\boldsymbol{\lambda}^*) \equiv [b_1 - g_1(\mathbf{x}^*), \ldots, b_m - g_m(\mathbf{x}^*)] \tag{2.12}$$

are non-negative, the next $(v-u)$ are non-positive and the rest vanish. Finally, each constraint is either active or passive, giving

Kuhn–Tucker condition IV (K–T IV)

$$[\nabla_\lambda F(\mathbf{x}^*,\boldsymbol{\lambda}^*)]'\boldsymbol{\lambda}^* \equiv \sum_i [b_i - g_i(\mathbf{x}^*)]\lambda_i{}^* = 0, \tag{2.13}$$

since either

$$b_i - g_i(\mathbf{x}^*) = 0 \quad (i \in I_a)$$

or

$$\lambda_i{}^* = 0 \quad (i \in I_p).$$

Example 2.1

For the following problem, verify that the Kuhn–Tucker necessary conditions are satisfied at the optimal point:

$$\textit{minimize} \quad z = x_1{}^2 - 2ax_1 + x_2 \quad (a > 0),$$

subject to

$$x_1 + 4x_2 \leqslant 2a,$$

$$x_1 + x_2 \geqslant a,$$
$$x_1, x_2 \geqslant 0.$$

Solution

Consider the equivalent problem of maximizing

$$-z = -x_1{}^2 + 2ax_1 - x_2 \qquad (a > 0).$$

Obviously, $\mathbf{x}^* = [a,0]$. The Lagrangian function is

$$F(\mathbf{x},\boldsymbol{\lambda}) \equiv -x_1{}^2 + 2ax_1 - x_2 + \lambda_1(2a - x_1 - 4x_2) + \lambda_2(a - x_1 - x_2).$$

Hence

$$\left.\begin{aligned} \frac{\partial F}{\partial x_1} &\equiv -2x_1 + 2a - \lambda_1 - \lambda_2, \\ \frac{\partial F}{\partial x_2} &\equiv -1 - 4\lambda_1 - \lambda_2, \\ \frac{\partial F}{\partial \lambda_1} &\equiv 2a - x_1 - 4x_2, \\ \frac{\partial F}{\partial \lambda_2} &\equiv a - x_1 - x_2. \end{aligned}\right\} \qquad (2.14)$$

Since the first constraint is passive, we have $\lambda_1{}^* = 0$. Also, from K–T I,

$$\frac{\partial F}{\partial x_1}(\mathbf{x}^*,\boldsymbol{\lambda}^*) = 0, \qquad (2.15)$$

which gives $\lambda_2{}^* = 0$. The optimal point $(\mathbf{x}^*,\boldsymbol{\lambda}^*)$ is therefore

$$[x_1{}^*, x_2{}^*, \lambda_1{}^*, \lambda_2{}^*] = [a,0,0,0], \qquad (2.16)$$

and from (2.14) we now find

$$\frac{\partial F}{\partial x_2}(\mathbf{x}^*,\boldsymbol{\lambda}^*) = -1, \quad \frac{\partial F}{\partial \lambda_1}(\mathbf{x}^*,\boldsymbol{\lambda}^*) = a, \quad \frac{\partial F}{\partial \lambda_2}(\mathbf{x}^*,\boldsymbol{\lambda}^*) = 0. \qquad (2.17)$$

It is easily verified that the values given by equations (2.15) to (2.17) satisfy K–T I to K–T IV.

It should be emphasized that the validity of the Kuhn–Tucker necessary conditions depends on the existence of the vector $\boldsymbol{\lambda}^*$. In order to ensure the existence of $\boldsymbol{\lambda}^*$, Kuhn and Tucker introduced a geometrical condition known as the *constraint qualification*; this is discussed more fully in Section 2.4. Meanwhile, we shall investigate the relationships between the Kuhn–Tucker necessary conditions and the saddle-point property of the Lagrangian function (2.9).

2.3 SADDLE-POINT PROPERTY OF THE LAGRANGIAN FUNCTION

Theorem 1.1 (page 20) shows that the Lagrangian function $F(\mathbf{x},\boldsymbol{\lambda})$ for the classical optimization problem (1.8) has a saddle-point at the optimal point $(\mathbf{x}^*, \lambda^*)$. In the present section, we shall consider the saddle-point properties of $F(\mathbf{x}, \lambda)$ for the general nonlinear programming problem (2.1). First, it is necessary to modify slightly the usual definition (1.61) of a saddle-point. We say that the Lagrangian function (2.9) for problem (2.1) has a saddle-point at $(\mathbf{x}^*, \lambda^*)$ if, in a neighbourhood of $(\mathbf{x}^*, \lambda^*)$,

$$F(\mathbf{x}, \lambda^*) \leqslant F(\mathbf{x}^*, \lambda^*) \leqslant F(\mathbf{x}^*, \lambda) \tag{2.18}$$

and

$$\begin{aligned} &\mathbf{x} \geqslant 0, \\ &\lambda_i \geqslant 0 \quad (i=1,....,u) \\ &\quad \leqslant 0 \quad (i=u+1,....,v) \\ &\text{is unrestricted in sign} \quad (i=v+1,....,m). \end{aligned} \tag{2.19}$$

Consider first the *necessary* conditions for a saddle-point of $F(\mathbf{x}, \lambda)$ at $(\mathbf{x}^*, \lambda^*)$. Suppose that $f, g_i \in C_1$ and that $F(\mathbf{x}, \lambda)$ has a saddle-point at $(\mathbf{x}^*, \lambda^*)$. Then, from the first inequality in (2.18), $F(\mathbf{x}, \lambda^*)$ takes on its maximum value with respect to $\mathbf{x}$ at the point $(\mathbf{x}^*, \lambda^*)$. Hence, for those values of j for which $x_j^* > 0$, we have from the ordinary theory of maxima and minima,

$$\frac{\partial F}{\partial x_j}(\mathbf{x}^*, \lambda^*) = 0 \quad (j \in J_p). \tag{2.20}$$

Similarly, from the second inequality in (2.18), $F(\mathbf{x}^*, \lambda)$ takes on its minimum value with respect to λ at $(\mathbf{x}^*, \lambda^*)$. Hence, for those components of λ that are unrestricted in sign, we have

$$\frac{\partial F}{\partial \lambda_i}(\mathbf{x}^*, \lambda^*) = 0 \quad (i=v+1,....,m); \tag{2.21}$$

also,

$$\frac{\partial F}{\partial \lambda_i}(\mathbf{x}^*, \lambda^*) = 0 \text{ (all values of } i=1,....,v \text{ for which } \lambda_i^* \neq 0). \tag{2.22}$$

Theorem 2.2 deals with the case where $x_j^* = 0$ for one or more values of j.

***Theorem* 2.2**

If, in (2.18), $x_j^* = 0$ *for some value of* j, *i.e. if* $j \in J_a$, *then* $\partial F(\mathbf{x}^*, \lambda^*)/\partial x_j \leqslant 0$.

Proof

The proof is by contradiction. Assume that $\partial F(\mathbf{x}^*, \lambda^*)/\partial x_j > 0$ for some value of $j \in J_a$. Then $\partial F(\mathbf{x}, \lambda)/\partial x_j > 0$ throughout a neighbourhood of $(\mathbf{x}^*, \lambda^*)$ in E^{n+m}, since $\partial F(\mathbf{x}, \lambda)/\partial x_j$ is continuous. Consider an expansion of $F(\mathbf{x}, \lambda^*)$ in this neighbourhood. Let

$$\mathbf{x} = \mathbf{x}^* + h\mathbf{e}_j \quad (h > 0),$$

where $\mathbf{e}_j$ is a unit vector in the positive x_j-direction. The condition $h > 0$ is imposed because of the non-negativity restriction $x_j \geqslant 0$. By the first mean value theorem,

$$F(\mathbf{x}^* + h\mathbf{e}_j, \boldsymbol{\lambda}^*) = F(\mathbf{x}^*, \boldsymbol{\lambda}^*) + h\frac{\partial F}{\partial x_j}(\mathbf{x}^* + \theta h\mathbf{e}_j, \boldsymbol{\lambda}^*), \quad 0 \leqslant \theta \leqslant 1, \tag{2.23}$$

$$> F(\mathbf{x}^*, \boldsymbol{\lambda}^*), \tag{2.24}$$

since both h and $\partial F/\partial x_j$ are strictly positive in equation (2.23). However, (2.24) contradicts the saddle-point property (2.18) of $F(\mathbf{x}, \boldsymbol{\lambda})$. Hence

$$\frac{\partial F}{\partial x_j}(\mathbf{x}^*, \boldsymbol{\lambda}^*) \leqslant 0 \text{ when } x_j^* = 0, \text{ i.e. when } j \in J_a. \tag{2.25}$$

Similar proofs to that of Theorem 2.2 show that

$$\frac{\partial F}{\partial \lambda_i}(\mathbf{x}^*, \boldsymbol{\lambda}^*) \geqslant 0 \quad (i = 1, \ldots., u), \tag{2.26}$$

$$\frac{\partial F}{\partial \lambda_i}(\mathbf{x}^*, \boldsymbol{\lambda}^*) \leqslant 0 \quad (i = u+1, \ldots., v). \tag{2.27}$$

Collecting together the results (2.20) to (2.22) and (2.25) to (2.27), and defining a saddle-point as in (2.18) and (2.19), we find that the necessary conditions for a saddle-point of $F(\mathbf{x}, \boldsymbol{\lambda})$ at the point $(\mathbf{x}^*, \boldsymbol{\lambda}^*)$ are equivalent to the Kuhn–Tucker necessary conditions (2.10) to (2.13) for problem (2.1). Specifically,

(2.20) and (2.25) $\Rightarrow$ (2.10) and (2.11),
(2.26), (2.27) and (2.21) $\Rightarrow$ (2.12),
(2.21) and (2.22) $\Rightarrow$ (2.13).

The saddle-point property of the Lagrangian function for the classical optimization problem (1.8) therefore extends, with the restrictions (2.19), to the more general nonlinear programming problem (2.1).

Next, we consider *sufficient* conditions for the Lagrangian function $F(\mathbf{x}, \boldsymbol{\lambda})$ of problem (2.1) to have a saddle-point at $(\mathbf{x}^*, \boldsymbol{\lambda}^*)$. The following theorem and its corollaries give the required results.

***Theorem* 2.3**

Let the point $(\mathbf{x}^*, \boldsymbol{\lambda}^*)$ *satisfy the necessary conditions,* (2.20) *to* (2.22) *and* (2.25) *to* (2.27), *for a saddle-point. Suppose that the components of* $\mathbf{x}$ *and* $\boldsymbol{\lambda}$ *satisfy* (2.19), *and suppose further that the points* $(\mathbf{x}, \boldsymbol{\lambda}^*)$ *and* $(\mathbf{x}^*, \boldsymbol{\lambda})$ *lie in a neighbourhood of* $(\mathbf{x}^*, \boldsymbol{\lambda}^*)$ *in which*

$$\text{(i)} \; F(\mathbf{x}, \boldsymbol{\lambda}^*) \leqslant F(\mathbf{x}^*, \boldsymbol{\lambda}^*) + (\mathbf{x} - \mathbf{x}^*)'\nabla_x F(\mathbf{x}^*, \boldsymbol{\lambda}^*), \tag{2.28}$$

$$\text{(ii)} \; F(\mathbf{x}^*, \boldsymbol{\lambda}) \geqslant F(\mathbf{x}^*, \boldsymbol{\lambda}^*) + (\boldsymbol{\lambda} - \boldsymbol{\lambda}^*)'\nabla_\lambda F(\mathbf{x}^*, \boldsymbol{\lambda}^*). \tag{2.29}$$

Then $F(\mathbf{x}, \boldsymbol{\lambda})$ *has a saddle-point at* $(\mathbf{x}^*, \boldsymbol{\lambda}^*)$.

Proof

Combining (2.20) and (2.25) gives K–T II:

$$\mathbf{x}^{*\prime}\nabla_x F(\mathbf{x}^*, \boldsymbol{\lambda}^*) = 0.$$

Hence

$$(\mathbf{x}-\mathbf{x}^*)'\nabla_x F(\mathbf{x}^*,\lambda^*) = \mathbf{x}'\nabla_x F(\mathbf{x}^*,\lambda^*). \tag{2.30}$$

Since $\mathbf{x} \geqslant 0$ by hypothesis, we can use (2.20) and (2.25) again to obtain

$$\mathbf{x}'\nabla_x F(\mathbf{x}^*,\lambda^*) \leqslant 0. \tag{2.31}$$

Then, from (2.28), (2.30) and (2.31), we find

$$F(\mathbf{x},\lambda^*) \leqslant F(\mathbf{x}^*,\lambda^*) + \mathbf{x}'\nabla_x F(\mathbf{x}^*,\lambda^*) \leqslant F(\mathbf{x}^*,\lambda^*). \tag{2.32}$$

Similarly, combining (2.21) and (2.22) gives K–T IV:

$$\lambda^{*\prime}\nabla_\lambda F(\mathbf{x}^*,\lambda^*) = 0.$$

Hence

$$(\lambda-\lambda^*)'\nabla_\lambda F(\mathbf{x}^*,\lambda^*) = \lambda'\nabla_\lambda F(\mathbf{x}^*,\lambda^*). \tag{2.33}$$

Since the components of λ satisfy (2.19), we can use (2.21), (2.22), (2.26) and (2.27) to obtain

$$\lambda'\nabla_\lambda F(\mathbf{x}^*,\lambda^*) \geqslant 0. \tag{2.34}$$

Then, from (2.29), (2.33) and (2.34), we find

$$F(\mathbf{x}^*,\lambda) \geqslant F(\mathbf{x}^*,\lambda^*) + \lambda'\nabla_\lambda F(\mathbf{x}^*,\lambda^*) \geqslant F(\mathbf{x}^*,\lambda^*). \tag{2.35}$$

Finally, combining (2.32) and (2.35), we obtain

$$F(\mathbf{x},\lambda^*) \leqslant F(\mathbf{x}^*,\lambda^*) \leqslant F(\mathbf{x}^*,\lambda),$$

which completes the proof of the theorem.

***Corollary* 1**

If (2.28) *and* (2.29) *hold for* all $\mathbf{x}$ *and* λ *satisfying* (2.19), *then* $F(\mathbf{x},\lambda)$ *has a* global *saddle-point at* $(\mathbf{x}^*,\lambda^*)$. (Conditions (2.18) and (2.19) define a *local* saddle-point.)

***Corollary* 2**

The conditions (2.28) *and* (2.29) *may be replaced, respectively, by*

(i) $F(\mathbf{x},\lambda^*)$ *is a concave function of* $\mathbf{x}$, (2.36)

(ii) $F(\mathbf{x}^*,\lambda)$ *is a convex function of* λ. (2.37)

Proof

In Theorem 1.3 (page 25), set

$$\mathbf{x}_1 = \lambda^*, \quad \mathbf{x}_2 = \lambda, \quad f(\mathbf{x}) = F(\mathbf{x}^*,\lambda).$$

Then (1.71) gives

$$(\lambda-\lambda^*)'\nabla_\lambda F(\mathbf{x}^*,\lambda^*) \leqslant F(\mathbf{x}^*,\lambda) - F(\mathbf{x}^*,\lambda^*),$$

which is identical with (2.29). Similarly, in Theorem 1.6 (page 26), set

$$\mathbf{x}_1 = \mathbf{x}^*, \quad \mathbf{x}_2 = \mathbf{x}, \quad f(\mathbf{x}) = F(\mathbf{x}, \boldsymbol{\lambda}^*).$$

Then (1.75) gives

$$(\mathbf{x} - \mathbf{x}^*)' \nabla_x F(\mathbf{x}^*, \boldsymbol{\lambda}^*) \geqslant F(\mathbf{x}, \boldsymbol{\lambda}^*) - F(\mathbf{x}^*, \boldsymbol{\lambda}^*),$$

which is identical with (2.28).

We conclude this section by obtaining sufficient conditions for problem (2.1) to have a global optimal solution. In fact, the proof of Theorem 2.4, which follows, shows that these conditions are identical with the sufficient conditions of Theorem 2.3, Corollary 1, for $F(\mathbf{x}, \boldsymbol{\lambda})$ to have a global saddle-point at $(\mathbf{x}^*, \boldsymbol{\lambda}^*)$.

***Theorem* 2.4**

In problem (2.1), *let the point* $(\mathbf{x}^*, \boldsymbol{\lambda}^*)$ *satisfy the Kuhn–Tucker necessary conditions,* (2.10) *to* (2.13), *and suppose that* $f(\mathbf{x})$ *is concave throughout the feasible region. Suppose also that the* $g_i(\mathbf{x})$ *for which* $\lambda_i^* > 0$ *are convex functions and that those for which* $\lambda_i^* < 0$ *are concave functions. Then* $f(\mathbf{x}^*)$ *is the constrained global maximum of* $f(\mathbf{x})$ *for problem* (2.1).

Proof

If the conditions of the theorem hold, then

$$F(\mathbf{x}, \boldsymbol{\lambda}^*) \equiv f(\mathbf{x}) + \sum_i \lambda_i^* [b_i - g_i(\mathbf{x})] \tag{2.38}$$

is a concave function of $\mathbf{x}$ for all $\mathbf{x} \geqslant \mathbf{0}$, since it is a sum of concave functions. Also, $F(\mathbf{x}^*, \boldsymbol{\lambda})$ is a convex function of $\boldsymbol{\lambda}$ for all $\boldsymbol{\lambda}$, since it is linear in $\boldsymbol{\lambda}$. Hence, by Corollaries 1 and 2 of Theorem 2.3, $F(\mathbf{x}, \boldsymbol{\lambda})$ has a global saddle-point at $(\mathbf{x}^*, \boldsymbol{\lambda}^*)$, i.e.

$$F(\mathbf{x}, \boldsymbol{\lambda}^*) \leqslant F(\mathbf{x}^*, \boldsymbol{\lambda}^*) \leqslant F(\mathbf{x}^*, \boldsymbol{\lambda}) \tag{2.39}$$

for all $\mathbf{x}$ and $\boldsymbol{\lambda}$ satisfying (2.19).

Setting $\mathbf{x} = \mathbf{x}^*$ in equation (2.38) and using equation (2.13), we find

$$F(\mathbf{x}^*, \boldsymbol{\lambda}^*) = f(\mathbf{x}^*). \tag{2.40}$$

Also, conditions (2.5) show that the signs of the λ_i^* are such that

$$\sum_i \lambda_i^* [b_i - g_i(\mathbf{x})] \geqslant 0. \tag{2.41}$$

It follows from (2.38) to (2.41) that for all $\mathbf{x}$ satisfying the constraints of problem (2.1),

$$f(\mathbf{x}) \leqslant F(\mathbf{x}, \boldsymbol{\lambda}^*) \leqslant F(\mathbf{x}^*, \boldsymbol{\lambda}^*) = f(\mathbf{x}^*),$$

which proves the theorem.

The main results of this section may be combined to give the following fundamental theorem, the statement of which should be carefully verified by the reader.

Theorem 2.5 The Kuhn–Tucker theorem

Let $f(\mathbf{x})$ and the $g_i(\mathbf{x})$ of the general nonlinear programming problem (2.1) *satisfy the convexity and concavity conditions of Theorem* 2.4. *Then the vector* $\mathbf{x}^*$ *is a global optimal solution of problem* (2.1) *if and only if a vector* λ^* *exists such that the Lagrangian function* (2.9) *has a global saddle-point at* $(\mathbf{x}^*, \lambda^*)$.

2.4 THE CONSTRAINT QUALIFICATION

It has already been indicated, in Section 2.2, that the Kuhn–Tucker necessary conditions, (2.10) to (2.13), are not always valid at an optimal point. A further condition is required in order to ensure the existence of a vector λ^* satisfying (2.10) to (2.13). In this section, we shall find a necessary and sufficient condition for the existence of the optimal Lagrange multipliers λ_i^* and λ_{m+j}^*. We shall then show how this condition is related to the constraint qualification, which is a condition on the active constraint boundaries near the optimal point designed to eliminate undesirable singularities such as outward-pointing cusps.

Consider first a two-dimensional example due to Kuhn and Tucker.[51]

Example 2.2

Maximize

$$z = x_1,$$

subject to

$$g_1(\mathbf{x}) \equiv (1 - x_1)^3 - x_2 \geqslant 0,$$
$$x_1, x_2 \geqslant 0.$$

Solution

The feasible region is shaded in Figure 2.1; it is obvious that $x_1^* = 1, x_2^* = 0$;

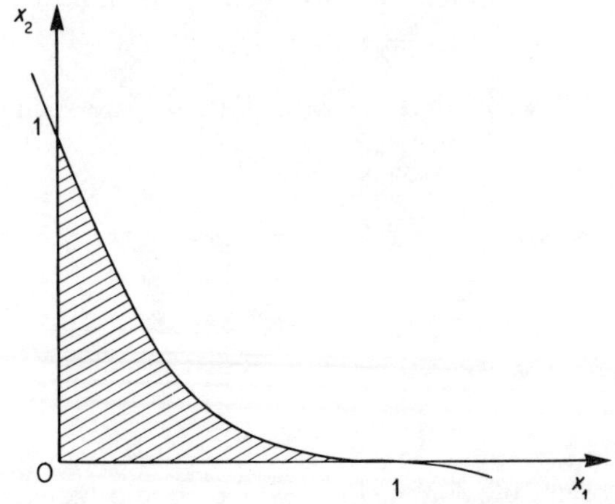

Figure 2.1 Feasible region for Example 2.2

$z^* = 1$. However, the Lagrangian function for this problem is

$$F(\mathbf{x},\lambda) \equiv x_1 + \lambda_1[x_2 - (1-x_1)^3],$$

and hence

$$\frac{\partial F}{\partial x_1}(\mathbf{x}^*,\lambda^*) \equiv 1 + 3\lambda_1{}^*(1-x_1{}^*)^2 = 1,$$

which violates the first Kuhn–Tucker necessary condition (2.10). The second Kuhn–Tucker necessary condition (2.11) is also violated. We shall return to this example later, to show that the constraint qualification is not satisfied at the point [1,0].

We now examine the conditions for the existence and uniqueness of the $\lambda_i{}^*$ and $\lambda_{m+j}{}^*$. Passive constraints and non-negativity restrictions need not be considered in the following analysis, since they remain passive in a sufficiently small neighbourhood of $\mathbf{x}^*$. Hence, we only consider constraints for which $i \in I_a$ and non-negativity restrictions for which $j \in J_a$.

In problem (2.1), suppose that m_a of the m constraints and n_a of the n non-negativity restrictions are active. Define the *active Lagrangian function*

$$F_a(\mathbf{x},\lambda) \equiv f(\mathbf{x}) + \sum_{i \in I_a} \lambda_i[b_i - g_i(\mathbf{x})] - \sum_{j \in J_a} \lambda_{m+j} x_j, \tag{2.42}$$

where the summations extend over the active constraints and non-negativity restrictions. In effect, the nonlinear programming problem (2.1) has been reduced to a classical optimization problem by ignoring the passive constraints and non-negativity restrictions. Hence, at the optimal point $(\mathbf{x}^*,\lambda^*)$, the λ_i and λ_{m+j} in (2.42) must satisfy

$$\frac{\partial F_a}{\partial x_j}(\mathbf{x}^*,\lambda^*) \equiv \frac{\partial f}{\partial x_j}(\mathbf{x}^*) - \sum_{i \in I_a} \lambda_i{}^* \frac{\partial g_i}{\partial x_j}(\mathbf{x}^*) - \lambda_{m+j}{}^* = 0 \quad (j \in J_a) \tag{2.43}$$

$$\equiv \frac{\partial f}{\partial x_j}(\mathbf{x}^*) - \sum_{i \in I_a} \lambda_i{}^* \frac{\partial g_i}{\partial x_j}(\mathbf{x}^*) = 0 \quad (j \in J_p). \tag{2.44}$$

Equations (2.43) and (2.44) may be combined to give the single vector equation

$$\sum_{i \in I_a} \lambda_i{}^* \nabla g_i(\mathbf{x}^*) + \sum_{j \in J_a} \lambda_{m+j}{}^* \mathbf{e}_j = \nabla f(\mathbf{x}^*), \tag{2.45}$$

where $\mathbf{e}_j$ is the unit coordinate vector in the x_j-direction.

Condition (2.45) states that $\nabla f(\mathbf{x}^*)$ belongs to the subspace of E^n spanned by the vectors $\nabla g_i(\mathbf{x}^*)$, $i \in I_a$, and $\mathbf{e}_j$, $j \in J_a$. Furthermore, the $\lambda_i{}^*$ must satisfy conditions (2.5) and the $\lambda_{m+j}{}^*$ must satisfy $\lambda_{m+j}{}^* \leqslant 0$ (see the proof of Theorem 2.1). All these conditions can be combined to give the single condition:

the vector $\nabla f(\mathbf{x}^*)$ must lie within or on the convex cone generated by the vectors

$$\begin{array}{lll} \nabla g_i(\mathbf{x}^*) & (i=1,\dots,u; & i\in I_a), \\ -\nabla g_i(\mathbf{x}^*) & (i=u+1,\dots,v; & i\in I_a), \\ -\mathbf{e}_j & (j\in J_a). & \end{array} \tag{2.46}$$

[A (convex) *cone* C is a (convex) set of points such that if $\mathbf{x}\in C$ then $\mu\mathbf{x}\in C$ for all $\mu \geqslant 0$. A convex cone is said to be *generated* by a finite set of vectors $\mathbf{x}_1,\dots,\mathbf{x}_r$ if it consists of the points $\mu\mathbf{x}$ for all $\mu \geqslant 0$, where $\mathbf{x}$ is any convex combination of $\mathbf{x}_1,\dots,\mathbf{x}_r$, i.e. where

$$\mathbf{x} = \sum_{k=1}^{r} \lambda_k \mathbf{x}_k, \text{ with } \sum_{k=1}^{r} \lambda_k = 1, \ \lambda_k \geqslant 0.]$$

Condition (2.46) is necessary and sufficient for the existence of Lagrange multipliers λ_i^* and λ_{m+j}^* for the maximizing problem (2.1). It follows that if (2.46) holds, then the Kuhn–Tucker necessary conditions are satisfiied—otherwise they are invalid.

When condition (2.46) is satisfied, the λ_i^* and λ_{m+j}^* in (2.45) are unique if and only if the vectors $\nabla g_i(\mathbf{x}^*)$, $i\in I_a$, and $\mathbf{e}_j$, $j\in J_a$, are linearly independent. Geometrically, these vectors are in the directions of the normals to the active constraint and coordinate boundaries, respectively, at the optimal point.

Returning to Example 2.2, the constraint $g_1(\mathbf{x})\geqslant 0$ and the non-negativity restriction $x_2 \geqslant 0$ are active at $\mathbf{x}^* = [1,0]$. Also,

$$\nabla f(\mathbf{x}^*) = [1,0], \ \nabla g_1(\mathbf{x}^*) = [0,-1], \ \mathbf{e}_2 = [0,1].$$

Hence $\nabla f(\mathbf{x}^*)$ does not belong to the subspace spanned by $\nabla g_1(\mathbf{x}^*)$ and $\mathbf{e}_2$, and so condition (2.45) is not satisfied. This explains why the Kuhn–Tucker necessary conditions are not satisfied at the optimal point $\mathbf{x}^*$ in Example 2.2. It is instructive to consider the following example, which differs from Example 2.2 only in the objective function.

Example 2.3

Minimize

$$z = x_2,$$

subject to

$$g_1(\mathbf{x}) \equiv (1-x_1)^3 - x_2 \geqslant 0,$$
$$x_1, x_2 \geqslant 0.$$

Solution

In accordance with our convention, we reformulate the problem as a maximizing problem:

$$\textit{maximize} \quad -z = -x_2.$$

It is obvious geometrically (Figure 2.1) that there are alternative optima, given by

$$0 \leqslant x_1^* \leqslant 1, \ x_2^* = 0; \ z^* = 0.$$

There are three cases to consider.

(i) $\mathbf{x}^* = [0,0]$.
$x_1 \geqslant 0$ and $x_2 \geqslant 0$ are active; $\mathbf{e}_1 = [1,0]$, $\mathbf{e}_2 = [0,1]$.

(ii) $\mathbf{x}^* = [x_1,0]$, $0 < x_1 < 1$.
$x_2 \geqslant 0$ is active; $\mathbf{e}_2 = [0,1]$.

(iii) $\mathbf{x}^* = [1,0]$.
$g_1(\mathbf{x}) \geqslant 0$ and $x_2 \geqslant 0$ are active; $\nabla g_1(\mathbf{x}^*) = [0,-1]$, $\mathbf{e}_2 = [0,1]$.

In each case and at every optimal point, including the point $[1,0]$, the vector $\nabla f(\mathbf{x}^*) = [0,-1]$ belongs to the subspace spanned by the 'active' subset of the vectors $\nabla g_1(\mathbf{x}^*)$, $\mathbf{e}_1$, $\mathbf{e}_2$, and condition (2.46) is satisfied.

It follows that the Kuhn–Tucker necessary conditions are satisfied at every optimal point, including the point $[1,0]$. This may be verified by forming the Lagrangian function

$$F(\mathbf{x},\boldsymbol{\lambda}) \equiv -x_2 + \lambda_1[x_2 - (1-x_1)^3]$$

and evaluating

$$\frac{\partial F}{\partial x_1}(\mathbf{x}^*,\boldsymbol{\lambda}^*) \equiv 3\lambda_1{}^*(1-x_1{}^*)^2$$

and

$$\frac{\partial F}{\partial x_2}(\mathbf{x}^*,\boldsymbol{\lambda}^*) \equiv -1 + \lambda_1{}^*.$$

If $x_1{}^* = 1$, then $\partial F(\mathbf{x}^*,\boldsymbol{\lambda}^*)/\partial x_1 = 0$; also, $\partial F(\mathbf{x}^*,\boldsymbol{\lambda}^*)/\partial x_2 < 0$, since $\lambda_1{}^* \leqslant 0$, by Theorem 2.1. If $x_1{}^* \neq 1$, then the constraint $g_1(\mathbf{x}) \geqslant 0$ is passive and hence $\lambda_1{}^* = 0$. Thus, for all $x_1{}^*$,

$$\frac{\partial F}{\partial x_1}(\mathbf{x}^*,\boldsymbol{\lambda}^*) = 0, \quad \frac{\partial F}{\partial x_2}(\mathbf{x}^*,\boldsymbol{\lambda}^*) < 0,$$

showing that K–T I is satisfied.

K–T II follows immediately from K–T I, since $\partial F(\mathbf{x}^*,\boldsymbol{\lambda}^*)/\partial x_1 = 0$ and $x_2{}^* = 0$.

K–T III is merely a restatement of the constraint $g_1(\mathbf{x}) \geqslant 0$ at the optimal point:

$$(1 - x_1{}^*)^3 - x_2{}^* \geqslant 0,$$

which is obviously satisfied by $0 \leqslant x_1{}^* \leqslant 1$, $x_2{}^* = 0$.

Finally, K–T IV states that

$$\lambda_1{}^*[x_2{}^* - (1-x_1{}^*)^3] = 0$$

and we have already shown that either $\lambda_1{}^* = 0$ or $[x_1{}^*,x_2{}^*] = [1,0]$.

The conclusion to be drawn from Examples 2.2 and 2.3 is that the validity of the Kuhn–Tucker necessary conditions depends on the form of the objective function as well as on the geometry of the constraint boundaries.

We shall now discuss the Kuhn–Tucker constraint qualification and its relationship with condition (2.46) for the existence of the Lagrange multipliers in the active Lagrangian function.

Consider the problem studied by Kuhn and Tucker:[51]

$$\left.\begin{aligned}&\text{maximize} \\ &\qquad z = f(\mathbf{x}), \\ &\text{subject to} \\ &\qquad \hat{g}_i(\mathbf{x}) \geqslant 0 \quad (i = 1, \dots, \hat{m}), \\ &\qquad \mathbf{x} \geqslant \mathbf{0}.\end{aligned}\right\} \tag{2.47}$$

The constraints in this problem are no less general than those of problem (2.1), and we shall subsequently examine the consequences of the constraint qualification for the latter problem.

Define the index set

$$\hat{I}_a = \{i : \hat{g}_i(\mathbf{x}^*) = 0\}.$$

In this analysis, as in the analysis leading to condition (2.46), it is only necessary to consider *active* constraints and non-negativity restrictions, i.e. those for which $i \in \hat{I}_a$ and $j \in J_a$, respectively. We assume that $\hat{m}_a$ of the $\hat{m}$ constraints and n_a of the n non-negativity restrictions are active.

Consider a point $\mathbf{x} = \mathbf{x}^* + \mathbf{dx}$, where the vector differential $\mathbf{dx}$ satisfies

$$\begin{aligned}[\nabla \hat{g}_i(\mathbf{x}^*)]'\mathbf{dx} &\geqslant 0 \qquad (i \in \hat{I}_a), \\ \mathbf{e}_j'\,\mathbf{dx} &\geqslant 0 \qquad (j \in J_a).\end{aligned} \tag{2.48}$$

Then the *constraint qualification* states that:

> to any $\mathbf{dx}$ satisfying (2.48), there corresponds a differentiable curve $\mathbf{x} = \mathbf{a}(\theta)$, $0 \leqslant \theta \leqslant 1$, satisfying the constraints and non-negativity restrictions of (2.47) at each point of its length, with the properties:
> (i) $\mathbf{x}^* = \mathbf{a}(0)$,
> (ii) there exists $\mu > 0$ such that $\mathrm{d}\mathbf{a}(0)/\mathrm{d}\theta = \mu \mathbf{dx}$.

Suppose that the constraint qualification is satisfied. Geometrically, $\mathbf{dx}$ is tangential to the curve $\mathbf{x} = \mathbf{a}(\theta)$ at the optimal point $\mathbf{x}^* = \mathbf{a}(0)$. Also, it can be shown by continuity arguments[44] that

$$[\nabla f(\mathbf{x}^*)]'\mathbf{dx} \leqslant 0. \tag{2.49}$$

Thus the constraint qualification ensures that the inequality (2.49) holds. If the constraint qualification is not satisfied then the inequality (2.49) may or may not hold.

To show the connection between the constraint qualification and the previous results of this section, we require a lemma, due to Farkas, on matrices and vectors.

Farkas' lemma

Suppose that $\mathbf{y}$ *satisfies* $\mathbf{G}\mathbf{y} \geqslant \mathbf{0}$. *Then* $\mathbf{b}$ *will satisfy* $\mathbf{b}'\mathbf{y} \geqslant 0$ *if and only if there exists a vector* $\mathbf{w} \geqslant \mathbf{0}$ *such that* $\mathbf{G}'\mathbf{w} = \mathbf{b}$.

Proof

Sufficient. Suppose $\mathbf{w} \geqslant \mathbf{0}$ exists such that $\mathbf{G}'\mathbf{w} = \mathbf{b}$. Then $\mathbf{w}'\mathbf{G}\mathbf{y} = \mathbf{b}'\mathbf{y}$ for any $\mathbf{y}$. Hence $\mathbf{b}'\mathbf{y} \geqslant 0$, since $\mathbf{G}\mathbf{y} \geqslant \mathbf{0}$.

Necessary. Suppose that $\mathbf{b}'\mathbf{y} \geqslant 0$ for every $\mathbf{y}$ satisfying $\mathbf{G}\mathbf{y} \geqslant \mathbf{0}$. Then the linear programming problem:

$$\text{minimize } z = \mathbf{b}'\mathbf{y}, \text{ subject to } \mathbf{G}\mathbf{y} \geqslant \mathbf{0},$$

has an optimal solution $\mathbf{y} = \mathbf{0}$. It follows from the fundamental theorem of duality in linear programming[75] that the dual problem, whose constraints and non-negativity restrictions are

$$\mathbf{G}'\mathbf{w} = \mathbf{b}, \quad \mathbf{w} \geqslant \mathbf{0},$$

also has an optimal solution, i.e. there exists a vector $\mathbf{w} \geqslant \mathbf{0}$ such that $\mathbf{G}'\mathbf{w} = \mathbf{b}$.

Note. In the above proof, every dual constraint is an *equality* constraint because every primal variable is unrestricted in sign.[75]

In Farkas' lemma, let

$$\mathbf{y} = \mathbf{dx}, \quad \mathbf{b} = -\nabla f(\mathbf{x}^*), \quad \mathbf{G} = \begin{pmatrix} \hat{\mathbf{G}}_a^* \\ \hline \mathbf{I}_r \end{pmatrix}, \tag{2.50}$$

where $\hat{\mathbf{G}}_a^*$ is the $(\hat{m}_a \times n)$ matrix whose ith row is $[\nabla \hat{g}_i(\mathbf{x}^*)]'$, $i \in \hat{I}_a$, and I_r is the $(n_a \times n)$ matrix which consists of the rows $j \in J_a$ of the unit matrix of order n. Also, let $\mathbf{w}$ be the $(\hat{m}_a + n_a)$-vector whose components are $-\lambda_i^*$, $i \in \hat{I}_a$, and $-\lambda_{\hat{m}+j}^*$, $j \in J_a$. Then, because of (2.48) and (2.49), Farkas' lemma states that there exist $\lambda_i^* \leqslant 0$, $i \in \hat{I}_a$, and $\lambda_{\hat{m}+j}^* \leqslant 0$, $j \in J_a$, such that

$$\sum_{i \in \hat{I}_a} \lambda_i^* \nabla \hat{g}_i(\mathbf{x}^*) + \sum_{j \in J_a} \lambda_{\hat{m}+j}^* \mathbf{e}_j = \nabla f(\mathbf{x}^*). \tag{2.51}$$

Since $\lambda_{\hat{m}+j}^* \leqslant 0$, $j \in J_a$, equation (2.51) implies

$$\nabla f(\mathbf{x}^*) - \sum_{i \in \hat{I}_a} \lambda_i^* \nabla \hat{g}_i(\mathbf{x}^*) \leqslant \mathbf{0}, \tag{2.52}$$

where the strict equality holds for components $j \in J_p$. It is important to note that the existence of the Lagrange multipliers λ_i^*, $i \in \hat{I}_a$, and $\lambda_{\hat{m}+j}^*$, $j \in J_a$, is already guaranteed by Farkas' lemma.

Geometrically, if we regard conditions (2.48) as fixing the direction of $\mathbf{dx}$, then (2.49) is a necessary condition on the direction of $\nabla f(\mathbf{x}^*)$. In fact, $\nabla f(\mathbf{x}^*)$ must lie within or on the convex cone generated by the vectors $-\nabla \hat{g}_i(\mathbf{x}^*)$, $i \in \hat{I}_a$, and $-\mathbf{e}_j$, $j \in J_a$; cf. condition (2.46). This follows immediately from equation (2.51), since $\lambda_i^* \leqslant 0$, $i \in \hat{I}_a$, and $\lambda_{\hat{m}+j}^* \leqslant 0$, $j \in J_a$.

The results we have just obtained for problem (2.47) can be applied to the original problem (2.1) by replacing the $(m-v)$ equality constraints

$$g_i(\mathbf{x}) = b_i \qquad (i = v+1, \ldots, m)$$

of problem (2.1) by the $2(m-v)$ active inequality constraints

$$g_i(\mathbf{x}) \leqslant b_i \qquad (i = v+1, \ldots, m),$$

$$g_i(\mathbf{x}) \geqslant b_i \qquad (i=v+1,....,m),$$

and then defining

$$\hat{g}_i(\mathbf{x}) \equiv b_i - g_i(\mathbf{x}) \qquad (i=1,....,\hat{u}) \tag{2.53}$$

$$\equiv g_i(\mathbf{x}) - b_i \qquad (i=\hat{u}+1,....,\hat{m}), \tag{2.54}$$

where $\hat{u} = u+m-v$, $\hat{m} = 2m-v$, so that $\hat{g}_i(\mathbf{x}) \geqslant 0$ for all i $(i=1,....,\hat{m})$. Substituting (2.53) and (2.54) into (2.51), we obtain once more an equation which is equivalent to (2.45), after identifying the λ_i^* $(i=1,....,u)$ in problem (2.1) with the negatives of the corresponding quantities in problem (2.47); this reversal of sign occurs because of the reversal of direction of the inequalities for $i=1,....,u$ between the two problems. The λ_i^* $(i=1,....,u)$ in problem (2.1) are then non-negative, in accordance with conditions (2.5). Corresponding to the *equality* constraints in problem (2.1), we have $(m-v)$ *pairs* of Lagrange multipliers $\pm\lambda_i^*$ in equation (2.51); this accounts for the slight changes in notation between equations (2.45) and (2.51).

Substituting (2.53) and (2.54) into (2.52), we obtain K–T I for problem (2.1), remembering that Lagrange multipliers for passive constraints take the value zero at the optimal point. The remaining Kuhn–Tucker necessary conditions, (2.11) to (2.13), are easily deduced, as before.

Since condition (2.46) follows immediately from equation (2.45), we have now shown that the constraint qualification implies K–T I to IV and also implies the existence of the optimal Lagrange multipliers. In other words, the constraint qualification guarantees the validity of the Kuhn–Tucker necessary conditions. As stated earlier, however, the constraint qualification, though sufficient, is not a necessary condition; it may fail to hold even though the optimal Lagrange multipliers do exist. In this case, the inequality (2.49) and the Kuhn–Tucker necessary conditions are satisfied. It is shown below that this situation arises at the optimal point $\mathbf{x}^* = [1,0]$ of example 2.3.

One important special case should be mentioned: if all the active constraints are linear then the constraint qualification is satisfied automatically, for the differentiable curve $\mathbf{x} = \mathbf{a}(\theta)$ can always be taken to be a straight line segment lying on the constraint boundary.

To end this section, consider again Examples 2.2 and 2.3. In both these examples, $\mathbf{x}^* = [1,0]$ is an optimal point and $\nabla g_1(\mathbf{x}^*) = [0,-1]$. From the first of the inequalities (2.48) we obtain $(-1)(\mathrm{d}x_2) \geqslant 0$. From the second of these inequalities we obtain $(+1)(\mathrm{d}x_2) \geqslant 0$, since $x_2 \geqslant 0$ is the only active non-negativity restriction. Hence $\mathrm{d}x_2 = 0$ and $\mathrm{d}x_1$ is unrestricted.

It is clear, however, from Figure 2.1 that if $\mathbf{dx}$ is taken in the positive x_1-direction at the optimal point $[1,0]$ it cannot be tangential to a curve which lies entirely *within* the feasible region. Hence the constraint qualification is not satisfied at this point. However, with this $\mathbf{dx}$, condition (2.49) is violated in Example 2.2 and is satisfied in Example 2.3. Correspondingly, the Kuhn–Tucker necessary conditions at the point $[1,0]$ are violated in Example 2.2 and are satisfied in Example 2.3. In Example 2.3, it is readily verified that the constraint qualification is satisfied at the remaining optimal points $x_1^* \in [0, 1)$, $x_2^* = 0$, and we have already seen (page 47) that the Kuhn–Tucker necessary conditions are satisfied at every optimal point.

2.5 QUADRATIC PROGRAMMING: WOLFE'S ALGORITHM

With the help of Kuhn–Tucker theory, we can derive Wolfe's algorithm[78] for the solution of a large class of quadratic programming problems. The general quadratic programming problem can be written:

maximize

$$z = \mathbf{x}'\mathbf{D}\mathbf{x} + \mathbf{c}'\mathbf{x},$$

subject to

$$\left.\begin{aligned} \mathbf{A}\mathbf{x} &\leqslant, = \text{ or } \geqslant \mathbf{b}, \\ \mathbf{x} &\geqslant 0. \end{aligned}\right\} \tag{2.55}$$

The matrix $\mathbf{D}$ is of order $(n \times n)$ and is assumed, without loss of generality, to be symmetric; $\mathbf{A}$ is an $(m \times n)$ matrix, $\mathbf{b}$ an m-component column vector and $\mathbf{c}'$ an n-component row vector. We assume that the quadratic form $\mathbf{x}'\mathbf{D}\mathbf{x}$ is either negative definite or negative semidefinite.

We now apply the Kuhn–Tucker necessary conditions, (2.10) to (2.13), to the problem. The Lagrangian function is

$$F(\mathbf{x}, \lambda) \equiv \mathbf{x}'\mathbf{D}\mathbf{x} + \mathbf{c}'\mathbf{x} + \lambda'(\mathbf{b} - \mathbf{A}\mathbf{x}).$$

From K–T I, we obtain

$$\nabla_x F(\mathbf{x}^*, \lambda^*) \equiv 2\mathbf{D}\mathbf{x}^* + \mathbf{c} - \mathbf{A}'\lambda^* \leqslant 0, \tag{2.56}$$

which is equivalent to

$$2\mathbf{D}\mathbf{x}^* + \mathbf{c} - \mathbf{A}'\lambda^* + \mathbf{v}^* = 0, \tag{2.57}$$

where $\mathbf{v}^* \geqslant \mathbf{0}$ is an n-component slack vector. From (2.56) and (2.57), we have

$$\mathbf{v}^* = -\nabla_x F(\mathbf{x}^*, \lambda^*),$$

and so K–T II gives

$$\mathbf{v}^{*\prime}\mathbf{x}^* = 0,$$

i.e.

$$v_j^* x_j^* = 0 \quad \text{for each value of} \quad j\,(j = 1, \ldots, n), \tag{2.58}$$

since non-negativity restrictions apply to both $\mathbf{v}^*$ and $\mathbf{x}^*$.

Condition K–T III is satisfied by any feasible solution and K–T IV may be expressed in the form

$$\lambda_i^* x_{n+i}^* = 0 \quad \text{for each value of} \quad i\,(i = 1, \ldots, v), \tag{2.59}$$

where x_{n+i} is the slack or surplus variable in the ith inequality constraint. Equations (2.59) express the obvious fact that each of the v *inequality* constraints of (2.55) is either active or passive. The vector of slack and surplus variables is denoted by

$$\mathbf{x}_s = [x_{n+1},....,x_{n+v}].$$

The signs of the λ_i* in equations (2.59) may be determined from Theorem 2.1. Specifically, if the constraints of problem (2.55) are

$$\left.\begin{aligned}\sum_j a_{ij}x_j &\leqslant b_i \qquad (i=1,....,u)\\ &\geqslant b_i \qquad (i=u+1,....,v)\\ &= b_i \qquad (i=v+1,....,m),\end{aligned}\right\} \tag{2.60}$$

then

$$\left.\begin{aligned}\lambda_i^* &\geqslant 0 \quad (i=1,....,u)\\ &\leqslant 0, \quad (i=u+1,....,v)\\ &\text{is unrestricted in sign} \quad (i=v+1,....,m).\end{aligned}\right\} \tag{2.61}$$

It follows from Theorem 2.4 that $\mathbf{x}^* = [x_1^*,....,x_n^*]$ is a global optimal solution of the quadratic programming problem (2.55) if we can find vectors $\mathbf{x}^* \geqslant 0$, $\mathbf{x}_s^* \geqslant 0$, $\boldsymbol{\lambda}^*$ and $\mathbf{v}^* \geqslant \mathbf{0}$ which satisfy the linear constraints

$$\left.\begin{aligned}\mathbf{Ax} &\leqslant, = \text{or} \geqslant \mathbf{b},\\ 2\mathbf{Dx} - \mathbf{A}'\boldsymbol{\lambda} + \mathbf{v} &= -\mathbf{c},\end{aligned}\right\} \tag{2.62}$$

together with the nonlinear conditions (2.58) and (2.59).

In numerical examples, it is convenient to follow the usual linear programming practice of using only non-negative variables. Let

$$\left.\begin{aligned}\lambda_i &= -\xi_i \qquad (i=u+1,....,v)\\ &= \zeta_i - \xi_i \qquad (i=v+1,....,m),\end{aligned}\right\} \tag{2.63}$$

where

$$\xi_i \geqslant 0 \ (i=u+1,....,m); \quad \zeta_i \geqslant 0 \ (i=v+1,....,m).$$

Then, because of (2.61), the constraints (2.62) become a set of linear constraints in which every variable may be assumed non-negative.

Expressing the first set of constraints in (2.62) in the form (2.60) and adding slack and surplus variables, we obtain the equations

$$\begin{aligned}\sum_j a_{ij}x_j + x_{n+i} &= b_i \qquad (i=1,....,u),\\ \sum_j a_{ij}x_j - x_{n+i} &= b_i \qquad (i=u+1,....,v),\\ \sum_j a_{ij}x_j \qquad &= b_i \qquad (i=v+1,....,m).\end{aligned} \tag{2.64}$$

The remaining constraints in (2.62) are already in the form of equations:

$$2\mathbf{Dx} - \mathbf{A}'\boldsymbol{\lambda} + \mathbf{v} = -\mathbf{c}. \tag{2.65}$$

The variables in equations (2.64) and (2.65) are $x_1,....,x_n$; $x_{n+1},....,x_{n+v}$; $\lambda_1,....,\lambda_m$; $v_1,....,v_n$; i.e. $(2n + m + v)$ in all. However, the nonlinear conditions (2.58) and (2.59) show that at least $(n + v)$ of these variables take the value zero. Hence there are at most $(n + m)$ variables which take non-zero values. Since $(n + m)$ is also the number of linear constraints in (2.62), it follows that any solution which satisfies the constraints (2.62) must be a *basic* solution of the linear equations (2.64) and (2.65).

It is therefore possible to use the simplex method to solve the quadratic programming problem (2.55). Phase 1 of the two-phase method[75] is used to obtain a basic solution of equations (2.64) and (2.65), subject to the nonlinear conditions (2.58) and (2.59), and satisfying $\mathbf{x} \geqslant \mathbf{0}$, $\mathbf{x}_s \geqslant \mathbf{0}$, $v \geqslant \mathbf{0}$, together with conditions (2.61) on the λ_i; see also equations (2.63). The only modification to the usual phase 1 procedure is that the variables x_j and v_j must never be allowed to appear together as *non-degenerate* basic variables for any value of j; similarly for the variables λ_i and x_{n+i} for any value of i $(i=1,....,v)$. We say that the simplex method is being used with *restricted basis entry*.

The following two examples illustrate the numerical procedures in detail.

Example 2.4

Maximize

$$z = -2x_1{}^2 - x_2{}^2 + 4x_1 + 6x_2,$$

subject to

$$\left.\begin{aligned} x_1 + 3x_2 &\leqslant 3, \\ x_1,x_2 &\geqslant 0. \end{aligned}\right\} \quad (2.66)$$

Solution

In the notation of (2.55), we have

$$\mathbf{A} = (1,3), \quad \mathbf{b} = [3], \quad \mathbf{D} = \begin{pmatrix} -2 & 0 \\ 0 & -1 \end{pmatrix}, \quad \mathbf{c} = [4,6].$$

Hence the constraints (2.62) become

$$\left.\begin{aligned} x_1 + 3x_2 \qquad\qquad &\leqslant 3, \\ -4x_1 \qquad - \lambda_1 + v_1 &= -4, \\ -2x_2 - 3\lambda_1 + v_2 &= -6. \end{aligned}\right\} \quad (2.67)$$

Since the constraint in (2.66) is of $\leqslant$ type, we may assume from (2.60) and (2.61) that $\lambda_1 \geqslant 0$, so that the substitutions of (2.63) are not needed. Adding a slack variable to the first constraint in (2.67), multiplying the second and third constraints throughout by -1 and adding artificial variables u_1 and u_2, we obtain

$$\left.\begin{aligned} x_1 + 3x_2 + x_3 \qquad\qquad\qquad\qquad &= 3, \\ 4x_1 \qquad + \lambda_1 - v_1 \qquad + u_1 \qquad &= 4, \\ 2x_2 \quad + 3\lambda_1 \qquad - v_2 \qquad + u_2 &= 6, \\ \text{all variables non-negative.} \end{aligned}\right\} \quad (2.68)$$

The nonlinear conditions (2.58) and (2.59) are

$$x_1 v_1 = 0, \quad x_2 v_2 = 0, \quad \lambda_1 x_3 = 0. \tag{2.69}$$

A basic feasible solution of equations (2.68) is

$$x_3 = 3, \quad u_1 = 4, \quad u_2 = 6,$$

and the phase 1 objective function to be *maximized* is

$$z_a = -u_1 - u_2. \tag{2.70}$$

The quadratic programming problem (2.66) has now been reduced to the linear programming problem given by (2.68) and (2.70), with the nonlinear conditions (2.69). The latter problem is solved by means of the following standard simplex tableaux:[75]

Tableau 1 (Example 2.4)

	$\mathbf{c}'$		0	0	0	0	0	0	−1	−1
$\mathbf{c}_B$	Basic Variables		x_1	x_2	x_3	λ_1	v_1	v_2	u_1	u_2
0	x_3	3	1	3	1	0	0	0	0	0
−1	u_1	4	4*	0	0	1	−1	0	1	0
−1	u_2	6	0	2	0	3	0	−1	0	1
	$z_j - c_j$	−10	−4	−2	0	−4	1	1	0	0

Tableau 2 (Example 2.4)

	$\mathbf{c}'$		0	0	0	0	0	0	−1
$\mathbf{c}_B$	Basic Variables		x_1	x_2	x_3	λ_1	v_1	v_2	u_2
0	x_3	2	0	3**	1	−1/4	1/4	0	0
0	x_1	1	1	0	0	1/4	−1/4	0	0
−1	u_2	6	0	2	0	3*	0	−1	1
	$z_j - c_j$	−6	0	−2	0	−3	0	1	0

According to the usual simplex rules, the pivot in Tableau 2 is the 3 in the λ_1 column; but this leads to a tableau in which both x_3 and λ_1 are non-degenerate variables, contradicting (2.69). We therefore choose the next best pivot, namely, the 3 in the x_2 column.

Tableau 4 is optimal. The optimal solution of the quadratic programming problem (2.66) is therefore

$$x_1^* = \frac{12}{19}, \quad x_2^* = \frac{15}{19}; \quad z^* = -\frac{288}{19^2} - \frac{225}{19^2} + \frac{48}{19} + \frac{90}{19} = \frac{111}{19}.$$

Tableau 3 (Example 2.4)

	c′		0	0	0	0	0	0	−1
$\mathbf{c}_B$	Basic Variables		x_1	x_2	x_3	λ_1	v_1	v_2	u_2
0	x_2	2/3	0	1	1/3	−1/12	1/12	0	0
0	x_1	1	1	0	0	1/4	−1/4	0	0
−1	u_2	14/3	0	0	−2/3	19/6*	−1/6	−1	1
	$z_j - c_j$	−14/3	0	0	2/3	−19/6	1/6	1	0

Tableau 4 (Example 2.4)

	c′		0	0	0	0	0	0
$\mathbf{c}_B$	Basic Variables		x_1	x_2	x_3	λ_1	v_1	v_2
0	x_2	15/19	0	1	6/19	0	3/38	−1/38
0	x_1	12/19	1	0	1/19	0	−9/38	3/38
0	λ_1	28/19	0	0	−4/19	1	−1/19	−6/19
	$z_j - c_j$	0	0	0	0	0	0	0

Note that $\lambda_1{}^* = \dfrac{28}{19} > 0$, in accordance with (2.61).

Example 2.5

Maximize

$$z = -x_1{}^2 + x_1 x_2 - 2x_2{}^2 + x_1 + x_2,$$

subject to

$$\left.\begin{aligned} x_1 - x_2 &\geqslant 3, \\ x_1 + x_2 &= 4. \end{aligned}\right\} \tag{2.71}$$

Solution

The simplest procedure in this example is to use the equality constraint to eliminate one of the variables. However, in order to illustrate Wolfe's algorithm in the case where equality constraints are present, we shall retain both variables in the calculation.

In the notation of (2.55), we have

$$\mathbf{A} = \begin{pmatrix} 1 & -1 \\ 1 & 1 \end{pmatrix}, \ \mathbf{b} = [3,4], \ \mathbf{D} = \begin{pmatrix} -1 & 1/2 \\ 1/2 & -2 \end{pmatrix}, \ \mathbf{c} = [1,1].$$

Note. The constraints associated with **A** are written down in the order $\leqslant, \geqslant, =$, as in (2.60).

The constraints (2.62) become

$$\begin{aligned}
x_1 - x_2 & \geq 3, \\
x_1 + x_2 & = 4, \\
-2x_1 + x_2 - \lambda_1 - \lambda_2 + v_1 & = -1, \\
x_1 - 4x_2 + \lambda_1 - \lambda_2 + v_2 & = -1.
\end{aligned}$$

From (2.60) and (2.61), we may assume that $\lambda_1 \leqslant 0$ and that λ_2 is unrestricted in sign. The substitutions (2.63) give

$$\begin{aligned}
\lambda_1 &= -\xi_1, \\
\lambda_2 &= \zeta_2 - \xi_2,
\end{aligned}$$

where $\xi_1 \geqslant 0$, $\zeta_2 \geqslant 0$, $\xi_2 \geqslant 0$. Subtracting a surplus variable from the first constraint, multiplying the third and fourth constraints throughout by -1 and adding an artificial variable to each constraint, we obtain

$$\left.\begin{aligned}
x_1 - x_2 - x_3 + u_1 &= 3, \\
x_1 + x_2 + u_2 &= 4, \\
2x_1 - x_2 - \xi_1 + \zeta_2 - \xi_2 - v_1 + u_3 &= 1, \\
-x_1 + 4x_2 + \xi_1 + \zeta_2 - \xi_2 - v_2 + u_4 &= 1, \\
\text{all variables non-negative.} &
\end{aligned}\right\} \quad (2.72)$$

The phase 1 linear programming problem is:

$$\text{maximize} \quad z_a = -u_1 - u_2 - u_3 - u_4,$$

subject to the constraints (2.72), with the nonlinear conditions

$$x_1 v_1 = 0, \; x_2 v_2 = 0, \; \xi_1 x_3 = 0. \qquad (2.73)$$

It is a straightforward matter to solve this problem by means of the simplex method, starting with the initial basic feasible solution

$$u_1 = 3, \; u_2 = 4, \; u_3 = 1, \; u_4 = 1.$$

However, in view of the large number of variables in equations (2.72), we shall use a slightly different method which is usually more efficient in this type of problem.

The first two constraints in (2.72) are written *without* artificial variables:

$$\begin{aligned}
x_1 - x_2 - x_3 &= 3, \\
x_1 + x_2 &= 4.
\end{aligned}$$

An obvious basic feasible solution of these equations is

$$x_1 = 4, \; x_3 = 1.$$

It is possible to use this basic feasible solution as part of the basic feasible solution for the constraints (2.72), provided we replace u_3 (which would otherwise be negative) by $-w_3$. The phase 1 linear programming problem becomes:

$$\text{maximize} \quad z_a = -w_3 - u_4,$$

subject to

$$\left.\begin{array}{rl} x_1 - x_2 - x_3 & = 3, \\ x_1 + x_2 & = 4, \\ 2x_1 - x_2 \quad - \xi_1 + \zeta_2 - \xi_2 - v_1 \quad - w_3 & = 1, \\ -x_1 + 4x_2 \quad + \xi_1 + \zeta_2 - \xi_2 \quad - v_2 \quad + u_4 & = 1, \\ \text{all variables non-negative,} & \end{array}\right\} \quad (2.74)$$

together with the nonlinear conditions (2.73).

A basic feasible solution of the constraint equations (2.74) is

$$x_1 = 4, \; x_3 = 1, \; w_3 = 7, \; u_4 = 5.$$

To construct the initial tableau, we solve the first two constraint equations of (2.74) for the basic variables x_1 and x_3 in terms of the non-basic variable x_2, giving

$$\begin{aligned} x_1 &= 4 - x_2, \\ x_3 &= 1 - 2x_2, \end{aligned}$$

or

$$\begin{aligned} x_1 + x_2 &= 4, \\ 2x_2 + x_3 &= 1. \end{aligned}$$

Substituting for x_1 and x_3 in the third and fourth equations of (2.74), we obtain

$$\begin{aligned} 3x_2 + \xi_1 - \zeta_2 + \xi_2 + v_1 \quad + w_3 \quad &= 7, \\ 5x_2 + \xi_1 + \zeta_2 - \xi_2 \quad - v_2 \quad + u_4 &= 5. \end{aligned}$$

The coefficients that appear in Tableau 1 are taken from the last four equations. The remaining tableaux are constructed by means of standard simplex transformations.

Tableau 4 is optimal. The optimal solution of the quadratic programming problem (2.71) is therefore

$$x_1{}^* = 7/2, \; x_2{}^* = 1/2; \; z^* = -7.$$

Tableau 1 (Example 2.5)

	$\mathbf{c}'$		0	0	0	0	0	0	0	0	−1	−1
$\mathbf{c}_B$	Basic Variables		x_1	x_2	x_3	ξ_1	ζ_2	ξ_2	v_1	v_2	w_3	u_4
0	x_1	4	1	1	0	0	0	0	0	0	0	0
0	x_3	1	0	2*	1	0	0	0	0	0	0	0
−1	w_3	7	0	3	0	1	−1	1	1	0	1	0
−1	u_4	5	0	5	0	1	1	−1	0	−1	0	1
	$z_j - c_j$	−12	0	−8	0	−2	0	0	−1	1	0	0

Tableau 2 (Example 2.5)

	$\mathbf{c}'$		0	0	0	0	0	0	0	0	-1	-1
$\mathbf{c}_B$	Basic Variables		x_1	x_2	x_3	ξ_1	ζ_2	ξ_2	v_1	v_2	w_3	u_4
0	x_1	7/2	1	0	$-1/2$	0	0	0	0	0	0	0
0	x_2	1/2	0	1	1/2	0	0	0	0	0	0	0
-1	w_3	11/2	0	0	$-3/2$	1	-1	1	1	0	1	0
-1	u_4	5/2	0	0	$-5/2$	1*	1	-1	0	-1	0	1
	$z_j - c_j$	-8	0	0	4	-2	0	0	-1	1	0	0

Tableau 3 (Example 2.5)

	$\mathbf{c}'$		0	0	0	0	0	0	0	0	-1
$\mathbf{c}_B$	Basic Variables		x_1	x_2	x_3	ξ_1	ζ_2	ξ_2	v_1	v_2	w_3
0	x_1	7/2	1	0	$-1/2$	0	0	0	0	0	0
0	x_2	1/2	0	1	1/2	0	0	0	0	0	0
-1	w_3	3	0	0	1	0	-2	2*	1	1	1
0	ξ_1	5/2	0	0	$-5/2$	1	1	-1	0	-1	0
	$z_j - c_j$	-3	0	0	-1	0	2	-2	-1	-1	0

Tableau 4 (Example 2.5)

	$\mathbf{c}'$		0	0	0	0	0	0	0	0
$\mathbf{c}_B$	Basic Variables		x_1	x_2	x_3	ξ_1	ζ_2	ξ_2	v_1	v_2
0	x_1	7/2	1	0	$-1/2$	0	0	0	0	0
0	x_2	1/2	0	1	1/2	0	0	0	0	0
0	ξ_2	3/2	0	0	1/2	0	-1	1	1/2	1/2
0	ξ_1	4	0	0	-2	1	0	0	1/2	$-1/2$
	$z_j - c_j$	0	0	0	0	0	0	0	0	0

Finally, we note that

$$\lambda_1{}^* = -\xi_1{}^* = -4 < 0, \quad \lambda_2{}^* = \zeta_2{}^* - \xi_2{}^* = -3/2 < 0,$$

in accordance with (2.61).

2.6 FURTHER DISCUSSION OF WOLFE'S ALGORITHM

Wolfe's algorithm for the solution of the quadratic programming problem relies for its effectiveness on the following theorem.

***Theorem* 2.6**

When $\mathbf{x}'\mathbf{D}\mathbf{x}$ *in the quadratic programming problem* (2.55) *is negative definite, the problem cannot have an unbounded solution.*

Proof

The objective function when $\mathbf{x} \neq \mathbf{0}$ is

$$z = \mathbf{x}'\mathbf{D}\mathbf{x}\left(1 + \frac{\mathbf{c}'\mathbf{x}}{\mathbf{x}'\mathbf{D}\mathbf{x}}\right). \tag{2.75}$$

Let $\mathbf{x}$ be any point on the hypersphere $|\mathbf{x}| = r$. Then $\mathbf{x} = r\hat{\mathbf{x}}$, where $\hat{\mathbf{x}}$ is a unit vector. Hence

$$\mathbf{x}'\mathbf{D}\mathbf{x} = r^2\hat{\mathbf{x}}'\mathbf{D}\hat{\mathbf{x}}.$$

Let m_0 be the maximum value of $\hat{\mathbf{x}}'\mathbf{D}\hat{\mathbf{x}}$. Then

$$\mathbf{x}'\mathbf{D}\mathbf{x} \leqslant r^2 m_0 < 0,$$

and therefore

$$\mathbf{x}'\mathbf{D}\mathbf{x} \to -\infty \text{ as } |\mathbf{x}| \to \infty. \tag{2.76}$$

Now let m_1 be the maximum value of $\left|\dfrac{\mathbf{c}'\hat{\mathbf{x}}}{\hat{\mathbf{x}}'\mathbf{D}\hat{\mathbf{x}}}\right|$. Then

$$\left|\frac{\mathbf{c}'\mathbf{x}}{\mathbf{x}'\mathbf{D}\mathbf{x}}\right| = \frac{1}{r}\left|\frac{\mathbf{c}'\hat{\mathbf{x}}}{\hat{\mathbf{x}}'\mathbf{D}\hat{\mathbf{x}}}\right| \leqslant \frac{m_1}{r},$$

and therefore

$$\frac{\mathbf{c}'\mathbf{x}}{\mathbf{x}'\mathbf{D}\mathbf{x}} \to 0 \text{ as } |\mathbf{x}| \to \infty. \tag{2.77}$$

It follows from (2.75) to (2.77) that

$$z \to -\infty \text{ as } |\mathbf{x}| \to \infty.$$

Thus z cannot become arbitrarily large in the maximizing problem (2.55), and the theorem is proved.

Theorem 2.6 does *not* hold when $\mathbf{x}'\mathbf{D}\mathbf{x}$ is negative semidefinite, although Wolfe's algorithm will often succeed in solving the problem in this case. If it does not, the elements of the matrix $\mathbf{D}$ can be perturbed slightly to become $\mathbf{D} + \varepsilon\mathbf{I}$, where $\varepsilon < 0$ and $|\varepsilon|$ is considerably less than the magnitude of any element of $\mathbf{D}$. Then $\mathbf{x}'(\mathbf{D} + \varepsilon\mathbf{I})\mathbf{x}$ is negative definite, since $\mathbf{x}'\mathbf{D}\mathbf{x} \leqslant 0$ for all $\mathbf{x}$, and $\varepsilon\mathbf{x}'\mathbf{I}\mathbf{x} < 0$ if $\mathbf{x} \neq \mathbf{0}$. Wolfe's algorithm will therefore succeed for the perturbed problem, and the numerical results will represent an acceptable solution of the unperturbed problem if $|\varepsilon|$ is sufficiently small.

A general constrained optimization problem, like a general linear programming problem, falls into one of the following three classes:

(a) There is no feasible solution.

(b) There is an unbounded solution.
(c) There is an optimal solution.

Combining this classification with Theorem 2.6 and Corollary 3 of Theorem 1.14 (page 31), we deduce that the quadratic programming problem (2.55) has a unique optimal solution if $\mathbf{x}'\mathbf{D}\mathbf{x}$ is negative definite and if a feasible solution exists. This means, in turn, that there must be a basic solution of the constraints (2.64) and (2.65), with $\mathbf{x} \geqslant \mathbf{0}$, $\mathbf{x}_s \geqslant \mathbf{0}$, $\mathbf{v} \geqslant \mathbf{0}$, the components of $\boldsymbol{\lambda}$ satisfying (2.61), and $\mathbf{x}, \mathbf{x}_s, \boldsymbol{\lambda}, \mathbf{v}$ satisfying the nonlinear conditions (2.58) and (2.59). It is shown by Hadley[44] that Wolfe's algorithm always leads to this basic solution, despite the nonlinear conditions (2.58) and (2.59).

Finally, we note that many algorithms have been proposed for the solution of the quadratic programming problem. Apart from Wolfe's algorithm, perhaps the best-known algorithm that depends on the simplex method is that of Beale.[4] Fletcher[35] has devised an algorithm that does not make use of the simplex method. Further details of several quadratic programming methods are given by Hadley.[44]

2.7 QUADRATIC PROGRAMMING AND DUALITY

It was shown in Theorem 1.2 that, under the assumptions (1.57), the dual of the problem:

$$\text{maximize} \quad f(\mathbf{x}), \quad \text{subject to } g_i(\mathbf{x}) = b_i,$$

is:

$$\text{minimize } F(\mathbf{x}, \boldsymbol{\lambda}), \quad \text{subject to } \frac{\partial F}{\partial x_j} = 0,$$

where

$$F(\mathbf{x}, \boldsymbol{\lambda}) \equiv f(\mathbf{x}) + \sum_i \lambda_i [b_i - g_i(\mathbf{x})].$$

First, we apply this result to the quadratic programming problem in which the only constraints are equality constraints.

Let the primal be the quadratic programming problem:

$$\left.\begin{aligned} &\text{maximize} \\ &\qquad z = \mathbf{x}'\mathbf{D}\mathbf{x} + \mathbf{c}'\mathbf{x}, \\ &\text{subject to} \\ &\qquad \mathbf{A}\mathbf{x} = \mathbf{b}, \\ &\qquad \mathbf{x} \text{ unrestricted in sign.} \end{aligned}\right\} \tag{2.78}$$

Then

$$F(\mathbf{x}, \boldsymbol{\lambda}) \equiv \mathbf{x}'\mathbf{D}\mathbf{x} + \mathbf{c}'\mathbf{x} + \boldsymbol{\lambda}'(\mathbf{b} - \mathbf{A}\mathbf{x}), \tag{2.79}$$

and the dual problem is:

minimize

$$\left.\begin{aligned} w &= \mathbf{x}'\mathbf{D}\mathbf{x} + \mathbf{c}'\mathbf{x} + \boldsymbol{\lambda}'(\mathbf{b} - \mathbf{A}\mathbf{x}), \\ &\text{subject to} \\ 2\mathbf{D}\mathbf{x} &+ \mathbf{c} - \mathbf{A}'\boldsymbol{\lambda} = \mathbf{0}. \end{aligned}\right\} \tag{2.80}$$

Problem (2.80) can be stated in the equivalent form:

minimize

$$w = -\mathbf{x}'\mathbf{D}\mathbf{x} + \mathbf{b}'\boldsymbol{\lambda},$$

subject to (2.81)

$$-2\mathbf{D}\mathbf{x} + \mathbf{A}'\boldsymbol{\lambda} = \mathbf{c}.$$

Problem (2.81) is a quadratic programming problem in the variables x_j, λ_i. Alternatively, it may be regarded as an unconstrained minimization problem in the λ_i, since the x_j can be eliminated by means of the constraint equation. Note that for problems (2.78) and (2.81),

$$w \geqslant w^* = F(\mathbf{x}^*, \boldsymbol{\lambda}^*) = z^* \geqslant z. \tag{2.82}$$

Next, we consider the dual of the general quadratic programming problem (2.55). The primal is:

maximize

$$\left.\begin{aligned} z &= \mathbf{x}'\mathbf{D}\mathbf{x} + \mathbf{c}'\mathbf{x}, \\ &\text{subject to} \\ \mathbf{A}\mathbf{x} &\leqslant, = \text{or} \geqslant \mathbf{b}, \\ \mathbf{x} &\geqslant \mathbf{0}. \end{aligned}\right\} \tag{2.83}$$

The constraints are written out explicitly in (2.60), and the Lagrangian function $F(\mathbf{x}, \boldsymbol{\lambda})$ is again given by (2.79).

Let $\mathbf{x}^*$ be the optimal solution of problem (2.83). Then

$$F(\mathbf{x}^*, \boldsymbol{\lambda}^*) = z^*,$$

and the components of $\boldsymbol{\lambda}^*$ satisfy conditions (2.61). Now for *any* vectors $\mathbf{x} \geqslant \mathbf{0}$ and $\boldsymbol{\lambda}$ satisfying

$$2\mathbf{D}\mathbf{x} + \mathbf{c} - \mathbf{A}'\boldsymbol{\lambda} \leqslant \mathbf{0} \quad [\text{cf. } (2.56)], \tag{2.84}$$

we have

$$-\boldsymbol{\lambda}'\mathbf{A}\mathbf{x} \leqslant -\mathbf{c}'\mathbf{x} - 2\mathbf{x}'\mathbf{D}\mathbf{x},$$

and so

$$\begin{aligned} F(\mathbf{x}, \boldsymbol{\lambda}) &\equiv \mathbf{x}'\mathbf{D}\mathbf{x} + \mathbf{c}'\mathbf{x} + \boldsymbol{\lambda}'\mathbf{b} - \boldsymbol{\lambda}'\mathbf{A}\mathbf{x} \\ &\leqslant -\mathbf{x}'\mathbf{D}\mathbf{x} + \boldsymbol{\lambda}'\mathbf{b} \\ &= w, \end{aligned} \tag{2.85}$$

the objective function in problem (2.81). However, K–T II for problem (2.83) gives

$$(2\mathbf{x}^{*\prime}\mathbf{D} + \mathbf{c}' - \boldsymbol{\lambda}^{*\prime}\mathbf{A})\mathbf{x}^* = 0,$$

and hence

$$\begin{aligned} z^* = F(\mathbf{x}^*, \boldsymbol{\lambda}^*) &= \mathbf{x}^{*\prime}\mathbf{D}\mathbf{x}^* + \mathbf{c}'\mathbf{x}^* + \boldsymbol{\lambda}^{*\prime}\mathbf{b} - \boldsymbol{\lambda}^{*\prime}\mathbf{A}\mathbf{x}^* \\ &= -\mathbf{x}^{*\prime}\mathbf{D}\mathbf{x}^* + \boldsymbol{\lambda}^{*\prime}\mathbf{b} \\ &= w^*. \end{aligned} \tag{2.86}$$

It follows from (2.84) to (2.86) that $(\mathbf{x}^*, \boldsymbol{\lambda}^*)$ is the optimal solution of the quadratic programming problem:

minimize

$$w = -\mathbf{x}'\mathbf{D}\mathbf{x} + \mathbf{b}'\boldsymbol{\lambda},$$

subject to

$$\left.\begin{aligned} -2\mathbf{D}\mathbf{x} + \mathbf{A}'\boldsymbol{\lambda} &\geqslant \mathbf{c}, \\ \mathbf{x} &\geqslant \mathbf{0}. \end{aligned}\right\} \tag{2.87}$$

Since the components of $\boldsymbol{\lambda}^*$ for this problem must satisfy conditions (2.61), we shall assume from now on that the components of $\boldsymbol{\lambda}$ in any solution, not necessarily optimal, also satisfy these conditions. Note that the relations (2.82) hold not only for problems (2.78) and (2.81) but also for problems (2.83) and (2.87).

Problem (2.87) is a quadratic programming problem in the $(n + m)$ variables x_j, λ_i, and is the dual of problem (2.83). We have shown that if (2.83) has an optimal solution $\mathbf{x}^*$, then (2.87) has an optimal solution $(\mathbf{x}^*, \boldsymbol{\lambda}^*)$. If $\mathbf{x}^*$ is known, then the remainder of the solution, represented by $\boldsymbol{\lambda}^*$, can be found by setting $\mathbf{x} = \mathbf{x}^*$ in (2.87) and solving the resulting *linear* programming problem. This follows from (2.86) and the saddle-point property, (2.18) and (2.19), of the Lagrangian function. In the pairs of dual problems (2.78), (2.81) and (2.83), (2.87), it should be noted that if z is a concave function of $\mathbf{x}$ then w is a convex function of $(\mathbf{x}, \boldsymbol{\lambda})$.

Example 2.6

Write down the dual of problem (2.71) *and solve it by the simplex method, using the optimal solution* $\mathbf{x}^*$ *of problem* (2.71).

Solution

Substituting the appropriate coefficients and constants in (2.87), we find that the dual of problem (2.71) is:

minimize

$$w = x_1{}^2 - x_1x_2 + 2x_2{}^2 + 3\lambda_1 + 4\lambda_2, \tag{2.88}$$

subject to

$$\begin{aligned} 2x_1 - x_2 + \lambda_1 + \lambda_2 &\geqslant 1, \\ -\,x_1 + 4x_2 - \lambda_1 + \lambda_2 &\geqslant 1, \\ x_1, x_2 &\geqslant 0, \\ \lambda_1 \leqslant 0, \quad \lambda_2 \text{ unrestricted in sign.} \end{aligned} \tag{2.88}$$

The optimal solution of problem (2.71) is

$$x_1{}^* = 7/2, \quad x_2{}^* = 1/2,$$

and so the linear programming problem for λ_1, λ_2 becomes:

$$\text{minimize} \quad w_1 = 3\lambda_1 + 4\lambda_2,$$

subject to

$$\begin{aligned} \frac{13}{2} + \lambda_1 + \lambda_2 &\geqslant 1, \\ -\frac{3}{2} - \lambda_1 + \lambda_2 &\geqslant 1, \\ \lambda_1 \leqslant 0, \quad \lambda_2 &\text{ unrestricted in sign,} \end{aligned}$$

where

$$w_1 = w - (x_1{}^{*2} - x_1{}^* x_2{}^* + 2x_2{}^{*2}).$$

The optimal solution of this problem is

$$\lambda_1{}^* = -4, \; \lambda_2{}^* = -3/2; \; w_1{}^* = -18.$$

The optimal solution of problem (2.88) is therefore

$$x_1{}^* = 7/2, \; x_2{}^* = 1/2, \; \lambda_1{}^* = -4, \; \lambda_2{}^* = -3/2; \; w^* = -7.$$

The justification for regarding (2.83) and (2.87) as dual problems is not only that the existence of an optimal solution $\mathbf{x}^*$ of (2.83) implies the existence of an optimal solution $(\mathbf{x}^*, \boldsymbol{\lambda}^*)$ of (2.87); the converse is also true, provided that $\mathbf{x}'\mathbf{D}\mathbf{x}$ is negative definite. To prove this, it is sufficient to show that the constraints of problem (2.83), namely,

$$\mathbf{A}\mathbf{x} \leqslant, = \text{or} \geqslant \mathbf{b},$$

have a feasible solution (see page 60).

It is convenient to reformulate problem (2.87) as a maximizing problem:

maximize

$$-w = \mathbf{x}'\mathbf{D}\mathbf{x} - \mathbf{b}'\boldsymbol{\lambda},$$

subject to

$$\left.\begin{aligned} 2\mathbf{D}\mathbf{x} - \mathbf{A}'\boldsymbol{\lambda} &\leqslant -\mathbf{c}, \\ \mathbf{x} &\geqslant \mathbf{0}, \\ \text{components of } \boldsymbol{\lambda} &\text{ satisfy (2.61).} \end{aligned}\right\} \tag{2.89}$$

The Lagrangian function for this problem is

$$\mathscr{F}(\mathbf{x},\boldsymbol{\lambda},\boldsymbol{\delta}) \equiv \mathbf{x}'\mathbf{D}\mathbf{x} - \mathbf{b}'\boldsymbol{\lambda} + \boldsymbol{\delta}'(-\mathbf{c} - 2\mathbf{D}\mathbf{x} + \mathbf{A}'\boldsymbol{\lambda}), \tag{2.90}$$

where the components of $\boldsymbol{\delta}$ are the Lagrange multipliers. Note carefully that $\mathbf{x}$ in the Lagrangian function (2.9) corresponds to both $\mathbf{x}$ and $\boldsymbol{\lambda}$ in (2.90), while $\boldsymbol{\lambda}$ in (2.9) becomes $\boldsymbol{\delta}$ in (2.90).

We wish to apply K–T I to problem (2.89). However, K–T I was derived under the assumption that all the variables in problem (2.1) satisfy non-negativity restrictions, whereas the variables of problem (2.89) include the components of $\boldsymbol{\lambda}$ and these satisfy conditions (2.61), i.e. some of the λ_i are non-negative, some non-positive and the rest are unrestricted in sign. It is easily verified that a more general form of K–T I is

$$\frac{\partial F}{\partial x_j}(\mathbf{x}^*,\boldsymbol{\delta}^*) \leqslant, = \text{or} \geqslant 0, \tag{2.91}$$

according as x_j is non-negative, unrestricted in sign or non-positive, where by x_j we now mean a component of either $\mathbf{x}$ or $\boldsymbol{\lambda}$ and the components of $\boldsymbol{\delta}$ are the Lagrange multipliers.

Let $(\mathbf{x}^*,\boldsymbol{\lambda}^*)$ be the optimal solution of problem (2.87). Applying the more general form (2.91) of K–T I to (2.90) and retaining only those components which involve derivatives with respect to the λ_i, we obtain

$$\frac{\partial}{\partial \lambda_i}\mathscr{F}(\mathbf{x}^*,\boldsymbol{\lambda}^*,\boldsymbol{\delta}^*) \equiv -b_i + \sum_j a_{ij}\,\delta_j^* \leqslant, = \text{or} \geqslant 0,$$

according as $\lambda_i^* \geqslant 0$, unrestricted in sign or $\leqslant 0$. That is,

$$\left.\begin{array}{lll} \sum_j a_{ij}\,\delta_j^* \leqslant b_i, & \lambda_i^* \geqslant 0 & (i=1,\ldots,u) \\ \qquad\qquad \geqslant b_i, & \lambda_i^* \leqslant 0 & (i=u+1,\ldots,v) \\ \qquad\qquad = b_i, & \lambda_i^* \text{ unrestricted in sign} & (i=v+1,\ldots,m). \end{array}\right\} \tag{2.92}$$

Furthermore, since the inequality constraints of problem (2.89) are of the form

$$g_j(\mathbf{x},\boldsymbol{\lambda}) \leqslant \text{ constant},$$

the corresponding Lagrange multipliers δ_j^* must be non-negative, by (2.61). Hence

$$\boldsymbol{\delta}^* \geqslant 0. \tag{2.93}$$

It follows from (2.92) and (2.93) that $\boldsymbol{\delta}^*$ is a feasible solution, as required, of the constraints of problem (2.83); see (2.60) for details of these constraints. This completes the proof that if problem (2.87) has an optimal solution then so has problem (2.83).

2.8 GRIFFITH AND STEWART'S METHOD

In 1961, Griffith and Stewart[42] extended the use of the simplex method to the general nonlinear programming problem:

maximize

$$\left.\begin{aligned} z &= f(\mathbf{x}), \\ &\text{subject to} \\ g_i(\mathbf{x}) &\leqslant, = \text{ or } \geqslant b_i, \\ \mathbf{x} &\geqslant 0. \end{aligned}\right\} \tag{2.94}$$

The basic idea of Griffith and Stewart's method is to expand each of the functions $f(\mathbf{x})$ and $g_i(\mathbf{x})$ about a feasible point $\mathbf{x}_1$ in a Taylor series which is truncated after the linear term. The nonlinear problem (2.94) is therefore approximated by a linear programming problem which can be solved to give a new trial point $\mathbf{x}_2$. The whole procedure is then repeated using $\mathbf{x}_2$ in place of $\mathbf{x}_1$. Thus the solution of problem (2.94) is reduced to the solution of a sequence of linear programming problems. Griffith and Stewart's method is also known as MAP (Method of Approximation Programming).

Let $\mathbf{x}_1$ be a feasible solution of problem (2.94). Then we write

$$\left.\begin{aligned} f(\mathbf{x}) &\doteqdot f(\mathbf{x}_1) + [\nabla f(\mathbf{x}_1)]'(\mathbf{x} - \mathbf{x}_1) \\ &\text{and} \\ g_i(\mathbf{x}) &\doteqdot g_i(\mathbf{x}_1) + [\nabla g_i(\mathbf{x}_1)]'(\mathbf{x} - \mathbf{x}_1). \end{aligned}\right\} \tag{2.95}$$

It is convenient to introduce new variables y_j, defined by

$$\mathbf{y} = [y_1, \ldots, y_n] = \mathbf{x} - \mathbf{x}_1. \tag{2.96}$$

Note that $\mathbf{y}$ is unrestricted in sign. In order to simplify the notation, we also define

$$\mathbf{a}_i = \nabla g_i(\mathbf{x}_1),\ \hat{b}_i = b_i - g_i(\mathbf{x}_1),\ \mathbf{c} = \nabla f(\mathbf{x}_1),\ \hat{z} = z - f(\mathbf{x}_1). \tag{2.97}$$

Substituting (2.95) into (2.94), and using (2.96) and (2.97), we obtain the linear programming problem:

maximize

$$\left.\begin{aligned} \hat{z} &= \mathbf{c}'\mathbf{y}, \\ &\text{subject to} \\ \mathbf{a}_i'\mathbf{y} &\leqslant, = \text{ or } \geqslant \hat{b}_i, \\ \mathbf{y} &\text{ unrestricted in sign.} \end{aligned}\right\} \tag{2.98}$$

Next, to ensure the validity of the linear approximations (2.95), we impose upper bounds on the magnitudes of the variables y_j:

$$|y_j| \leqslant m_j. \tag{2.99}$$

Let $\mathbf{y}_1^*$ be the optimal solution of problem (2.98), subject to the additional constraints (2.99). Then

$$\mathbf{x}_2 = \mathbf{x}_1 + \mathbf{y}_1^*$$

is taken as the next trial point, the constants $\mathbf{a}_i$, $\hat{b}_i$, $\mathbf{c}$ of (2.97) are evaluated at this point and a new linear programming problem, similar to (2.98) with conditions (2.99), is formulated.

Even when conditions (2.99) are included, there is no guarantee that the new trial point $\mathbf{x}_2$ will satisfy the constraints of problem (2.94). If it does not, we can either decrease the upper bounds m_j and re-solve the linear programming problem (2.98) or proceed to the next approximation ignoring the fact that $\mathbf{x}_2$ is not a feasible solution of problem (2.94).

The iterations terminate when the difference between two successive solutions is acceptably small, e.g. when

$$|\mathbf{x}_{r+1} - \mathbf{x}_r| < \varepsilon, \tag{2.100}$$

where $\varepsilon > 0$ is prescribed. As an alternative to (2.100), or in addition to it, we can use a condition on successive values of the objective function:

$$|z_{r+1} - z_r| < \delta,$$

where $\delta > 0$ is prescribed. Unfortunately, convergence cannot be guaranteed; nevertheless, the method has been used successfully on many practical problems.

The choice of the m_j in (2.99) involves the usual compromise associated with step lengths in iterative processes: accuracy is lost if the m_j are made too large, while an excessive number of iterations is required if they are made too small. The m_j can, of course, be adjusted as the calculations proceed.

In many practical problems the variables x_j are bounded. Suppose that

$$x_j' \leqslant x_j \leqslant x_j''. \tag{2.101}$$

Since $\mathbf{y}$ is unrestricted in sign in problem (2.98) but is bounded in magnitude by (2.99), Beale[4] suggests that the simplest procedure is to replace the variables y_j by the non-negative variables

$$w_j = y_j + m_j. \tag{2.102}$$

Then (2.101) can be written

$$x_j' \leqslant x_{1j} + w_j - m_j \leqslant x_j'', \tag{2.103}$$

where x_{1j} is the jth component of $\mathbf{x}_1$. Also, (2.99) becomes

$$-m_j \leqslant w_j - m_j \leqslant m_j. \tag{2.104}$$

Both (2.103) and (2.104) are satisfied if

$$\max\{x_j' - x_{1j} + m_j, 0\} \leqslant w_j \leqslant \min\{x_j'' - x_{1j} + m_j, 2m_j\}. \tag{2.105}$$

Substituting equations (2.102) into problem (2.98), we find that the linear programming problem to be solved at each iteration is:

maximize

$$\left.\begin{aligned} &\hat{z} = \mathbf{c}'(\mathbf{w} - \mathbf{m}), \\ &\text{subject to} \\ &\mathbf{a}_i'\mathbf{w} \leqslant, = \text{ or } \geqslant \hat{b}_i + \mathbf{a}_i'\mathbf{m}, \\ &w_j' \leqslant w_j \leqslant w_j'', \end{aligned}\right\} \tag{2.106}$$

where

$$\begin{aligned} \mathbf{w} &= [w_1,\ldots,w_n],\ \mathbf{m} = [m_1,\ldots,m_n], \\ w_j' &= \max\{x_j' - x_{1j} + m_j, 0\}, \\ w_j'' &= \min\{x_j'' - x_{1j} + m_j, 2m_j\}, \end{aligned}$$

and x_{1j} refers to the current trial point. The expressions for w_j' and w_j'' may also be used when, instead of (2.101), the only bounds on the x_j are non-negativity restrictions. In this case, $x_j' = 0$ and $x_j'' \to \infty$. The constant term $-\mathbf{c}'\mathbf{m}$ may be omitted from the objective function.

If $\mathbf{w}_r^*$ is the optimal solution of problem (2.106) in the rth iteration (which begins with the trial point $\mathbf{x}_r$), then the next trial point is

$$\mathbf{x}_{r+1} = \mathbf{x}_r + \mathbf{y}_r^* = \mathbf{x}_r + \mathbf{w}_r^* - \mathbf{m}. \tag{2.107}$$

Example 2.7

Maximize

$$\left.\begin{aligned} &z = 2x_1^2 - x_1x_2 + 3x_2^2, \\ &\textit{subject to} \\ &3x_1 + 4x_2 \leqslant 12, \\ &x_1^2 - x_2^2 \geqslant 1, \\ &x_1, x_2 \geqslant 0. \end{aligned}\right\} \tag{2.108}$$

Solution

Take $\mathbf{x}_1 = [2,1]$ as the initial feasible solution. From equations (2.97), we obtain

$$\begin{aligned} &\mathbf{a}_1 = [3,4],\ \mathbf{a}_2 = [2x_{11}, -2x_{12}] = [4,-2], \\ &\hat{b}_1 = 2,\ \hat{b}_2 = -2,\ \mathbf{c} = [4x_{11} - x_{12}, -x_{11} + 6x_{12}] = [7,4]. \end{aligned}$$

In problem (2.108), the first constraint and the non-negativity restrictions imply

$$0 \leqslant x_1 \leqslant 4, \quad 0 \leqslant x_2 \leqslant 3.$$

We may therefore take

$$x_1' = 0,\ x_1'' = 4,\ x_2' = 0,\ x_2'' = 3$$

as the upper and lower bounds in (2.103). Throughout the calculation we shall take $m_1 = m_2 = \frac{1}{2}$ as the maximum allowable change in each component of $\mathbf{x}_1, \mathbf{x}_2, \ldots$. Thus $\mathbf{m} = [\frac{1}{2}, \frac{1}{2}]$ and the linear programming problem (2.106) becomes:

maximize

$$\left.\begin{aligned} \hat{z} &= 7w_1 + 4w_2, \\ \text{subject to} \\ 3w_1 + 4w_2 &\leqslant \frac{11}{2}, \\ 4w_1 - 2w_2 &\geqslant -1, \\ 0 \leqslant w_1 \leqslant 1,\ & 0 \leqslant w_2 \leqslant 1. \end{aligned}\right\} \tag{2.109}$$

The constant term $-\mathbf{c}'\mathbf{m}$ has been omitted from the objective function $\hat{z}$ and will be omitted from subsequent objective functions.

The optimal solution of problem (2.109) is $\mathbf{w}_1{}^* = [1, \frac{5}{8}]$. Hence (2.107) gives

$$\mathbf{x}_2 = \mathbf{x}_1 + \mathbf{w}_1{}^* - \mathbf{m} = [2,1] + \left[1, \frac{5}{8}\right] - \left[\frac{1}{2}, \frac{1}{2}\right] = \left[\frac{5}{2}, \frac{9}{8}\right].$$

This completes the first iteration.

Replacing $\mathbf{x}_1$ by $\mathbf{x}_2$ in equations (2.97), we obtain

$$\mathbf{a}_1 = [3,4],\ \mathbf{a}_2 = \left[5, -\frac{9}{4}\right],\ \hat{b}_1 = 0,\ \hat{b}_2 = -\frac{255}{64},\ \mathbf{c} = \left[\frac{71}{8}, \frac{17}{4}\right].$$

Problem (2.106) becomes:

maximize

$$\left.\begin{aligned} \hat{z} &= \frac{71}{8}w_1 + \frac{17}{4}w_2, \\ \text{subject to} \\ 3w_1 + 4w_2 &\leqslant \frac{7}{2}, \\ 5w_1 - \frac{9}{4}w_2 &\geqslant -\frac{167}{64}, \\ 0 \leqslant w_1 \leqslant 1,\ & 0 \leqslant w_2 \leqslant 1. \end{aligned}\right\}$$

We find $\mathbf{w}_2{}^* = [1, \frac{1}{8}]$ and

$$\mathbf{x}_3 = \mathbf{x}_2 + \mathbf{w}_2{}^* - \mathbf{m} = \left[\frac{5}{2}, \frac{9}{8}\right] + \left[1, \frac{1}{8}\right] - \left[\frac{1}{2}, \frac{1}{2}\right] = \left[3, \frac{3}{4}\right],$$

which completes the second iteration.

Replacing $\mathbf{x}_1$ by $\mathbf{x}_3$ in equations (2.97), we obtain

$$\mathbf{a}_1 = [3,4],\ \mathbf{a}_2 = \left[6, -\frac{3}{2}\right],\ \hat{b}_1 = 0,\ \hat{b}_2 = -\frac{119}{16},\ \mathbf{c} = \left[\frac{45}{4}, \frac{3}{2}\right].$$

Problem (2.106) becomes:

maximize

$$\left.\begin{aligned}\hat{z} &= \frac{45}{4}w_1 + \frac{3}{2}w_2,\\ &\text{subject to}\\ 3w_1 + 4w_2 &\leqslant \frac{7}{2},\\ 6w_1 - \frac{3}{2}w_2 &\geqslant -\frac{83}{16},\\ 0 \leqslant w_1 \leqslant 1,&\ 0 \leqslant w_2 \leqslant 1.\end{aligned}\right\}$$

We find $\mathbf{w}_3{}^* = [1, \frac{1}{8}]$ and

$$\mathbf{x}_4 = \mathbf{x}_3 + \mathbf{w}_3{}^* - \mathbf{m} = \left[3, \frac{3}{4}\right] + \left[1, \frac{1}{8}\right] - \left[\frac{1}{2}, \frac{1}{2}\right] = \left[\frac{7}{2}, \frac{3}{8}\right],$$

which completes the third iteration.

Replacing $\mathbf{x}_1$ by $\mathbf{x}_4$ in equations (2.97), we obtain

$$\mathbf{a}_1 = [3,4],\ \mathbf{a}_2 = \left[7, -\frac{3}{4}\right],\ \hat{b}_1 = 0,\ \hat{b}_2 = -\frac{711}{64},\ \mathbf{c} = \left[\frac{109}{8}, -\frac{5}{4}\right].$$

Problem (2.106) becomes:

maximize

$$\left.\begin{aligned}\hat{z} &= \frac{109}{8}w_1 - \frac{5}{4}w_2,\\ &\text{subject to}\\ 3w_1 + 4w_2 &\leqslant \frac{7}{2},\\ 7w_1 - \frac{3}{4}w_2 &\geqslant -\frac{511}{64},\\ 0 \leqslant w_1 \leqslant 1,&\quad \frac{1}{8} \leqslant w_2 \leqslant 1.\end{aligned}\right\}$$

We find $\mathbf{w}_4{}^* = [1, \frac{1}{8}]$, and

$$\mathbf{x}_5 = \mathbf{x}_4 + \mathbf{w}_4{}^* - \mathbf{m} = \left[\frac{7}{2}, \frac{3}{8}\right] + \left[1, \frac{1}{8}\right] - \left[\frac{1}{2}, \frac{1}{2}\right] = [4,0],$$

which completes the fourth iteration.

Replacing $\mathbf{x}_1$ by $\mathbf{x}_5$ in equations (2.97), we obtain

$$\mathbf{a}_1 = [3,4],\ \mathbf{a}_2 = [8,0],\ \hat{b}_1 = 0,\ \hat{b}_2 = -15,\ \mathbf{c} = [16, -4].$$

Problem (2.106) becomes:

maximize

$$\left.\begin{array}{l} \hat{z} = 16w_1 - 4w_2, \\ \text{subject to} \\ 3w_1 + 4w_2 \leqslant \frac{7}{2}, \\ 8w_1 \geqslant -11, \\ 0 \leqslant w_1 \leqslant \frac{1}{2},\ \frac{1}{2} \leqslant w_2 \leqslant 1. \end{array}\right\}$$

We find $\mathbf{w}_5^* = [\frac{1}{2}, \frac{1}{2}]$, and so

$$\mathbf{y}_5 = \mathbf{w}_5^* - \mathbf{m} = \mathbf{0},$$

showing that we have reached the *exact* optimal solution of problem (2.108):

$$\mathbf{x}^* = \mathbf{x}_5 = [4,0]; \ z^* = 32.$$

Note. (i) The values of the objective function z in successive iterations are

$$z = 9,\ 863/64,\ 279/16,\ 1511/64,\ 32;$$

these show a steady improvement.

(ii) The values of x_1 in successive iterations are

$$x_1 = 2,\ 5/2,\ 3,\ 7/2,\ 4.$$

This arithmetic progression shows that the number of iterations that were needed to reach the optimal solution is the smallest possible with $m_1 = \frac{1}{2}$.

(iii) It is instructive to solve problem (2.108) graphically and to plot the path by which Griffith and Stewart's method reaches the optimal point. The effect of increasing the value of m_1 after the first iteration should also be investigated.

SUMMARY

Kuhn–Tucker theory is an extension of the theory of classical optimization to the case where inequality constraints are present. The fundamental theorem on the subject is the Kuhn–Tucker theorem (Theorem 2.5, page 44), which gives necessary and sufficient conditions for a point to be globally optimal. This theorem is derived by first finding necessary conditions for a local optimum, adding convexity and concavity requirements to make them sufficient for a global optimum, and then showing that the sufficient conditions are equivalent to the condition that the Lagrangian function has a global saddle-point at the point in question. Condition (2.46) is necessary and sufficient for the existence of the Lagrange multipliers in Kuhn–Tucker theory. The constraint qualification is a geometrical condition which is sufficient, though not necessary, for the existence of these multipliers. Sections 2.5 to 2.7 deal with the quadratic programming problem and, in particular, Wolfe's algorithm for its solution.

The algorithm is a simple deduction from Kuhn–Tucker theory, though its practical implementation requires some knowledge of the simplex method for linear programming. Finally, Griffith and Stewart's method is described in Section 2.8. The basic idea of this method is to proceed iteratively by linearizing the objective function and the constraint functions about the current point, thereby reducing the given nonlinear problem to a sequence of linear programming problems.

EXERCISES

1. Verify that the Kuhn–Tucker necessary conditions are satisfied at the optimal point for the following problem:

$$\text{maximize} \quad z = x_1{}^2 - x_2,$$

subject to

$$\begin{aligned} x_1 + 3x_2 &\leqslant 6, \\ x_1 + x_2 &\geqslant 3, \\ x_1, x_2 &\geqslant 0. \end{aligned}$$

Are the sufficient conditions for a constrained global maximum (Theorem 2.4) satisfied?

2. Obtain the Kuhn–Tucker necessary conditions for the following problem:

$$\text{minimize} \quad z = f(\mathbf{x}),$$

subject to

$$\begin{aligned} &g_i(\mathbf{x}) \leqslant 0 \quad (i = 1, \ldots., m), \\ &\mathbf{a} \leqslant \mathbf{x} \leqslant \mathbf{b}. \end{aligned}$$

3. In Example 2.3, is there a unique value of the optimal Lagrange multiplier $\lambda_1{}^*$ at the point $\mathbf{x}^* = [1,0]$? Explain.

4. Investigate the effect of applying the Kuhn–Tucker necessary conditions to the linear programming problem:

$$\text{maximize} \quad z = \mathbf{c}'\mathbf{x},$$

subject to

$$\mathbf{A}\mathbf{x} \leqslant \mathbf{b}, \ \mathbf{x} \geqslant \mathbf{0}.$$

5. Consider the nonlinear programming problem:

$$\text{maximize} \quad z = x_1{}^2 + x_2{}^2,$$

subject to

$$\begin{aligned} (x_2 - b)^2 &\geqslant 4ax_1 \qquad (0 < b < 4a), \\ x_2 &\leqslant b, \\ x_1, x_2 &\geqslant 0. \end{aligned}$$

Verify that the following conditions are satisfied at the optimal point.

(a) The Kuhn–Tucker necessary conditions.

(b) The constraint qualification.

(c) Condition (2.46).

(d) Condition (2.49).

6. Consider the nonlinear programming problem:

$$\text{minimize} \quad z = (x_1 - a)^2 + (x_2 - a)^2,$$

subject to

$${x_1}^{2/5} + {x_2}^{2/5} \leqslant a^{2/5}.$$

Are the Kuhn–Tucker necessary conditions satisfied at the optimal points? If not, why not?

7. Consider the three nonlinear programming problems:

(i) maximize x_2,

(ii) maximize $x_2 - x_1$,

(iii) minimize x_1,

subject in each case to the constraints

$$x_1 \leqslant (1 - x_2)^5,$$
$$x_1, x_2 \geqslant 0.$$

Which of the conditions (a) to (d) of Exercise 5, if any, are satisfied at the optimal points?

8. Suppose that $\mathbf{y}$ satisfies $\mathbf{Gy} \geqslant \mathbf{0}$. Prove that $\mathbf{b}$ will satisfy $\mathbf{b'y} \geqslant 0$ for all $\mathbf{y} \geqslant \mathbf{0}$ if and only if $\mathbf{w} \geqslant \mathbf{0}$ exists such that $\mathbf{G'w} \leqslant \mathbf{b}$.

What is the effect of using the substitutions (2.50) to relate this variant of Farkas' lemma to the constraint qualification?

[Apply Wolfe's algorithm to Exercises 9 to 13.]

9. Maximize $z = -{x_1}^2 + x_1x_2 - 2{x_2}^2 + x_1 + x_2$, subject to

$$2x_1 + x_2 \leqslant 1,$$
$$x_1, x_2 \geqslant 0.$$

10. Maximize $z = -{x_1}^2 + 2x_1x_2 - 2{x_2}^2 + 2x_1 + 5x_2$, subject to

$$2x_1 + 3x_2 \leqslant 20,$$
$$3x_1 - 5x_2 \leqslant 4,$$
$$x_1, x_2 \geqslant 0.$$

11. Maximize $z = -{x_1}^2 + 3x_1x_2 - 3{x_2}^2 + 2x_1 + 5x_2$, subject to

$$x_1 + 4x_2 \leqslant 7,$$
$$3x_1 + x_2 = 4.$$
$$x_1, x_2 \geqslant 0.$$

12. Maximize $z = {x_1}^2 - 2x_1 - x_2$, subject to

$$x_1 + 5x_2 \leqslant 12,$$
$$x_1 + x_2 \geqslant 2,$$
$$x_1, x_2 \geqslant 0.$$

13. Maximize $z = -3x_1^2 + 2x_1x_2 - 2x_2^2 + 2x_1 + 3x_2$, subject to

$$\begin{aligned} x_1 + \ x_2 &\geqslant 1, \\ 3x_1 + 4x_2 &\leqslant 12, \\ x_1, x_2 &\geqslant 0. \end{aligned}$$

14. Write down and solve the dual of the problem of Exercise 13, using the optimal solution of the primal. Comment on the result.

15. Regard the linear programming problem of Exercise 4 as a special case of a quadratic programming problem. Show that its dual is the usual dual of a linear programming problem.

16. Solve the problem of Exercise 12 by Griffith and Stewart's method.

17. Perform one complete iteration of Griffith and Stewart's method on the following problem, using $\mathbf{x}_1 = [2,2,2]$ as the initial point and $\mathbf{m} = [1,1,1]$ as the vector of maximum allowable step lengths:

$$\text{maximize } z = x_1x_2x_3,$$

subject to

$$\begin{aligned} x_1(x_2 + x_3) &\leqslant 10, \\ e^{x_3}(\sin x_1 + \sin x_2) &\leqslant 30, \\ x_1, x_2, x_3 &\geqslant 0. \end{aligned}$$

[It is convenient to use the notation $p = e^2 \cos 2$, $q = e^2 \sin 2$.]

CHAPTER 3

Search Methods for Unconstrained Optimization

3.1 INTRODUCTION

In this chapter, we shall consider the problem of maximizing or minimizing a function $f(\mathbf{x})$ of n real variables x_j under the assumption that no constraint is imposed on the values of the x_j. In practice, this assumption is not a serious limitation; many methods for dealing with constraints have been devised, and we shall see in Chapter 5 that most of these methods can be used in conjunction with any technique for unconstrained optimization.

A *direct search method* is a method which relies only on evaluating $f(\mathbf{x})$ at a sequence of points $\mathbf{x}_1, \mathbf{x}_2, \ldots$ and comparing values, in order to reach the optimal point $\mathbf{x}^*$. Direct search methods are commonly used in the following circumstances:

(a) The function $f(\mathbf{x})$ is not differentiable, or is subject to random error.

(b) The derivatives $\partial f / \partial x_j$ are discontinuous, or their evaluation is much more difficult than the evaluation of the function $f(\mathbf{x})$ itself.

(c) Computer facilities are inadequate for the more complicated algorithms.

(d) Insufficient time is available in which to obtain a satisfactory result by other methods.

(e) An approximate solution may be required at any time during the course of the calculations.

The fundamental problem in the design of a direct search method is to determine the point $\mathbf{x}_{r+1}$, given the points $\mathbf{x}_1, \ldots, \mathbf{x}_r$ and the function values $f(\mathbf{x}_1), \ldots, f(\mathbf{x}_r)$. In Sections 3.2 to 3.7 we consider some well-known direct search methods, roughly in order of increasing complexity.

In Sections 3.8 and 3.9 we discuss the important subject of *linear search techniques*, i.e. methods for finding a maximum or minimum of $f(\mathbf{x})$ along a given line; this is essentially a one-dimensional problem. Some linear search techniques involve the evaluation of the derivatives of $f(\mathbf{x})$, others do not; both types are in common use. Linear searches are used in conjunction with many gradient methods, but further discussion of this useful application is deferred until Chapter 4.

3.2 GRID SEARCH

Probably the simplest direct search method is to construct a rectangular grid of points, evaluate $f(\mathbf{x})$ at each of these points in turn and take the greatest (or least) value of $f(\mathbf{x})$ so obtained as $f(\mathbf{x}^*)$. If we know that $\mathbf{x}^*$ lies within the region defined by

$$x_j' \leqslant x_j \leqslant x_j'',$$

and if each interval

$$\delta_j = x_j'' - x_j'$$

is divided into m_j sub-intervals by $(m_j + 1)$ points, then the total number of function evaluations required is

$$\prod_{j=1}^{n} (m_j + 1).$$

Two major disadvantages of this scheme are:

(a) It can only be used in a small region near the optimal point, otherwise a large number of unwanted function values are calculated.

(b) Information about $f(\mathbf{x})$ that is being acquired during the calculations is not being used subsequently to speed up the search for the optimal point.

Instead of evaluating $f(\mathbf{x})$ at every point of the grid, a *random search* is somctimes carried out over a smaller number of points; a table of random numbers may be used to choose these points. However, for a given number of function evaluations, a random search is usually less efficient in locating the optimal point than a full search over a coarser grid.

The second disadvantage mentioned above can be overcome to some extent by the following feedback strategy:

1. Choose a point $\mathbf{x}_1$. Evaluate $f(\mathbf{x})$ at $\mathbf{x} = \mathbf{x}_1$ and at the points immediately surrounding it, i.e. at the points $\mathbf{x}_1 + \sum_j h_j \mathbf{e}_j$, where the $\mathbf{e}_j$ are the unit coordinate vectors and each h_j takes the three values $0, \pm\delta_j/m_j$. Thus, in n dimensions, there are 3^n such points, including $\mathbf{x}_1$. The quantity δ_j/m_j is the grid spacing, assumed constant, on the x_j-axis.

2. Let $\mathbf{x}_2$ be the point which gives the greatest (least) value of $f(\mathbf{x})$ in step 1. If $\mathbf{x}_2 = \mathbf{x}_1$, reduce the grid spacing and repeat step 1. Otherwise, repeat step 1 using $\mathbf{x}_2$ in place of $\mathbf{x}_1$.

3. Continue until $\mathbf{x}^*$ is obtained within the desired accuracy.

Although this method is considerably more efficient than a comprehensive grid search, the number of function evaluations required in each iteration is still prohibitive unless n is fairly small.

3.3 HOOKE AND JEEVES' METHOD

Hooke and Jeeves' method,[46] which dates from 1961, is one of the most widely used direct search methods. It attempts in a simple though ingenious way to find the most profitable search directions. Constraints can easily be taken into account—see Exercise 4 at the end of this chapter. We shall consider the problem of minimizing $f(\mathbf{x})$; it is easy to write down the corresponding statements for a maximizing problem.

We choose an initial *base point* $\mathbf{b}_1$ and step lengths h_j for the respective variables x_j. For greater numerical accuracy it is advisable to choose the h_j so as to equalize, as far as possible, the quantities

$$|f(\mathbf{b}_1 + h_j\mathbf{e}_j) - f(\mathbf{b}_1)|;$$

these are the magnitudes of the changes in $f(\mathbf{b}_1)$ due to a change of one step length in each variable in turn. After $f(\mathbf{b}_1)$ has been evaluated, the method proceeds by a sequence of *exploratory* and *pattern moves.* If an exploratory move leads to a decrease in the value of $f(\mathbf{x})$ it is called a *success*; otherwise it is called a *failure.* A pattern move is not tested for success or failure.

Exploratory moves

The purpose of an exploratory move is to acquire information about $f(\mathbf{x})$ in the neighbourhood of the current base point. (One is reminded of Dougal of the Magic Roundabout who looks around in all directions before moving off!) The procedure for an exploratory move about the point $\mathbf{b}_1$ is as follows:

E(i) Evaluate $f(\mathbf{b}_1 + h_1\mathbf{e}_1)$. If the move from $\mathbf{b}_1$ to $\mathbf{b}_1 + h_1\mathbf{e}_1$ is a success, replace the base point $\mathbf{b}_1$ by $\mathbf{b}_1 + h_1\mathbf{e}_1$. If it is a failure, evaluate $f(\mathbf{b}_1 - h_1\mathbf{e}_1)$. If this move is a success, replace $\mathbf{b}_1$ by $\mathbf{b}_1 - h_1\mathbf{e}_1$. If it is another failure, retain the original base point $\mathbf{b}_1$.

E(ii) Repeat E(i) for the variable x_2 by considering variations $\pm h_2\mathbf{e}_2$ from the point which results from E(i). Apply this procedure to each variable in turn, finally arriving at a new base point $\mathbf{b}_2$ after $(2n + 1)$ function evaluations at most, including $f(\mathbf{b}_1)$.

E(iii) If $\mathbf{b}_2 = \mathbf{b}_1$, halve each of the step lengths h_j and return to E(i). The calculations terminate when the step lengths have been reduced to some prescribed level. If $\mathbf{b}_2 \neq \mathbf{b}_1$, make a pattern move from $\mathbf{b}_2$.

Pattern moves and subsequent moves

A pattern, or leapfrog, move attempts to speed up the search by using information already acquired about $f(\mathbf{x})$. It is invariably followed by a sequence of exploratory moves, with a view to finding an improved direction of search in which to make another pattern move. We shall denote by $\mathbf{p}_1, \mathbf{p}_2, \ldots$ the points reached by successive pattern moves. It seems sensible to move from $\mathbf{b}_2$ in the direction $(\mathbf{b}_2 - \mathbf{b}_1)$, since a move in this direction has already led to a decrease in the value of $f(\mathbf{x})$. The procedure for a pattern move from $\mathbf{b}_2$ is therefore as follows:

P(i) Move from $\mathbf{b}_2$ to $\mathbf{p}_1 = 2\mathbf{b}_2 - \mathbf{b}_1$ and continue with a new sequence of exploratory moves about $\mathbf{p}_1$.

P(ii) If the lowest function value obtained during the pattern and exploratory moves of P(i) is less than $f(\mathbf{b}_2)$, then a new base point $\mathbf{b}_3$ has been reached. In this case, return to P(i) with all suffices increased by unity. Otherwise, abandon the pattern move from $\mathbf{b}_2$ and continue with a new sequence of exploratory moves about $\mathbf{b}_2$.

Hooke and Jeeves in their excellent paper[46] report that their method has been used successfully to solve curve-fitting problems for which other methods failed; also, large systems of ill-conditioned linear equations have been solved economically by this method. In each case, the problem was reformulated as an optimization problem.

Example 3.1

Use Hooke and Jeeves' method to minimize

$$f(\mathbf{x}) \equiv 3x_1^{\,2} - 2x_1x_2 + x_2^{\,2} + 4x_1 + 3x_2.$$

Take $\mathbf{b}_1 = [0,0]$ *as the initial base point,* $h_1 = h_2 = 1$ *as initial step lengths and* $h_1 = h_2 < \frac{1}{4}$ *as the stopping condition.*

Solution

Let $\mathrm{E}(\mathbf{x}_r)$ denote a sequence of exploratory moves about the point $\mathbf{x}_r$, let $\mathrm{P}(\mathbf{b}_r)$ denote a pattern move from the base point $\mathbf{b}_r$, and let S and F denote a success and a failure, respectively, of an exploratory move relative to $\mathbf{x}_r$.
First, $f(\mathbf{b}_1) = 0$.

$\mathrm{E}(\mathbf{b}_1)$	$f(1,0) = 7$	(F)
	$f(-1,0) = -1$	(S)
	$f(-1,1) = 5$	(F)
	$f(-1,-1) = -5$	(S)

Since $\mathrm{E}(\mathbf{b}_1)$ is a success, the new base point is

$$\mathbf{b}_2 = [-1,-1];\ f(\mathbf{b}_2) = -5.$$

$\mathrm{P}(\mathbf{b}_2)$	$\mathbf{p}_1 = 2\mathbf{b}_2 - \mathbf{b}_1 = [-2,-2];\ f(\mathbf{p}_1) = -6.$	
$\mathrm{E}(\mathbf{p}_1)$	$f(-1,-2) = -7$	(S)
	$f(-1,-1) = -5$	(F)
	$f(-1,-3) = -7$	(F)

From $\mathrm{P}(\mathbf{b}_2)$ and $\mathrm{E}(\mathbf{p}_1)$, the new base point is

$$\mathbf{b}_3 = [-1,-2];\ f(\mathbf{b}_3) = -7,$$

and we make a further pattern move.

$\mathrm{P}(\mathbf{b}_3)$ $\qquad \mathbf{p}_2 = 2\mathbf{b}_3 - \mathbf{b}_2 = [-1,-3];\ f(\mathbf{p}_2) = -7.$

$E(\mathbf{p}_2)$	$f(0, -3) = 0$ (F)
	$f(-2, -3) = -8$ (S)
	$f(-2, -2) = -6$ (F)
	$f(-2, -4) = -8$ (F)

From $P(\mathbf{b}_3)$ and $E(\mathbf{p}_2)$, the new base point is

$$\mathbf{b}_4 = [-2, -3];\ f(\mathbf{b}_4) = -8,$$

and we make a further pattern move.

$P(\mathbf{b}_4)$	$\mathbf{p}_3 = 2\mathbf{b}_4 - \mathbf{b}_3 = [-3, -4];\ f(\mathbf{p}_3) = -5.$
$E(\mathbf{p}_3)$	$f(-2, -4) = -8$ (S)
	$f(-2, -3) = -8$ (F)
	$f(-2, -5) = -6$ (F)

Since $P(\mathbf{b}_4)$ and $E(\mathbf{p}_3)$ still leave $f(\mathbf{b}_4)$ as the lowest function value, we abandon $P(\mathbf{b}_4)$ and continue with exploratory moves about $\mathbf{b}_4$.

$E(\mathbf{b}_4)$	$f(-1, -3) = -7$ (F)
	$f(-3, -3) = -3$ (F)
	$f(-2, -2) = -6$ (F)
	$f(-2, -4) = -8$ (F)

Since $E(\mathbf{b}_4)$ is a failure, we reduce the step lengths to $h_1 = h_2 = \frac{1}{2}$.

$E(\mathbf{b}_4)$	$f(-3/2, -3) = -33/4$ (S)
	$f(-3/2, -5/2) = -8$ (F)
	$f(-3/2, -7/2) = -8$ (F)

Since $E(\mathbf{b}_4)$ is a success, the new base point is

$$\mathbf{b}_5 = [-3/2, -3];\ f(\mathbf{b}_5) = -33/4.$$

$P(\mathbf{b}_5)$	$\mathbf{p}_4 = 2\mathbf{b}_5 - \mathbf{b}_4 = [-1, -3];\ f(\mathbf{p}_4) = -7.$
$E(\mathbf{p}_4)$	$f(-1/2, -3) = -17/4$ (F)
	$f(-3/2, -3) = -33/4$ (S)
	$f(-3/2, -5/2) = -8$ (F)
	$f(-3/2, -7/2) = -8$ (F)

Since $P(\mathbf{b}_5)$ and $E(\mathbf{p}_4)$ still leave $f(\mathbf{b}_5)$ as the lowest function value, we abandon $P(\mathbf{b}_5)$ and continue with exploratory moves about $\mathbf{b}_5$.

$E(\mathbf{b}_5)$	$f(-1, -3) = -7$ (F)
	$f(-2, -3) = -8$ (F)
	$f(-3/2, -5/2) = -8$ (F)
	$f(-3/2, -7/2) = -8$ (F)

Since $E(\mathbf{b}_5)$ is a failure, we reduce the step lengths to $h_1 = h_2 = \frac{1}{4}$.

$E(\mathbf{b}_5)$	$f(-5/4, -3) = -125/16$ (F)
	$f(-7/4, -3) = -133/16$ (S)
	$f(-7/4, -11/4) = -65/8$ (F)

$$f(-7/4, -13/4) = -67/8 \text{ (S)}$$

Since E($\mathbf{b}_5$) is a success, the new base point is

$$\mathbf{b}_6 = [-7/4, -13/4]; \; f(\mathbf{b}_6) = -67/8.$$

P($\mathbf{b}_6$) $\quad \mathbf{p}_5 = 2\mathbf{b}_6 - \mathbf{b}_5 = [-2, -7/2]; \; f(\mathbf{p}_5) = -33/4.$

E($\mathbf{p}_5$) $\quad f(-7/4, -7/2) = -133/16 \quad$ (S)

$\quad f(-7/4, -13/4) = -67/8 \quad$ (S)

Since P($\mathbf{b}_6$) and E($\mathbf{p}_5$) still leave $f(\mathbf{b}_6)$ as the lowest function value, we abandon P($\mathbf{b}_6$) and continue with exploratory moves about $\mathbf{b}_6$.

E($\mathbf{b}_6$) $\quad f(-3/2, -13/4) = -131/16 \quad$ (F)

$\quad f(-2, -13/4) = -131/16 \quad$ (F)

$\quad f(-7/4, -3) = -133/16 \quad$ (F)

$\quad f(-7/4, -7/2) = -133/16 \quad$ (F)

Since E($\mathbf{b}_6$) is a failure, we reduce the step lengths to $h_1 = h_2 = \frac{1}{8}$. This, however, is the prescribed stopping condition and so the final result (which happens to be exact) is

$$\mathbf{x}^* = \mathbf{b}_6 = [-7/4, -13/4]; \; f(\mathbf{x}^*) = -67/8.$$

3.4 SPENDLEY, HEXT AND HIMSWORTH'S METHOD

We have already noted that it is desirable in any direct search method to make full use of the function evaluations already available. With this precept in mind, Spendley, Hext and Himsworth[73] devised in 1962 a method based on the geometrical design known as a *regular simplex*. A *simplex* in E^n consists of $(n+1)$ points $\mathbf{x}_t$ $(t = 1, \ldots, n+1)$ which do not lie on a hyperplane, together with every convex combination of these points. In the present application, we shall be concerned only with the $(n+1)$ points $\mathbf{x}_t$, which are called the *vertices* of the simplex. The simplex is *regular* if its vertices are equally spaced. Examples of regular simplices are an equilateral triangle in E^2 and a regular tetrahedron in E^3. The basic ideas of Spendley, Hext and Himsworth's method will be explained for the two-dimensional case; the extension to n dimensions is straightforward.

Two-dimensional case

Consider the problem of minimizing $f(\mathbf{x})$. Let $\mathbf{x}_1$ be an initial estimate of $\mathbf{x}^*$ and let $\mathbf{x}_1, \mathbf{x}_2, \mathbf{x}_3$ be the vertices of a regular simplex in E^2.

The *reflection* (Figure 3.1) of the vertex $\mathbf{x}_1$ in the line segment joining $\mathbf{x}_2$ and $\mathbf{x}_3$ is defined in a natural way as

$$\mathbf{x}_4 = \mathbf{x}_2 + \mathbf{x}_3 - \mathbf{x}_1. \tag{3.1}$$

It is obvious that the points $\mathbf{x}_2, \mathbf{x}_3, \mathbf{x}_4$ are also the vertices of a regular simplex. Similarly, if $\mathbf{x}_2$ or $\mathbf{x}_3$ is reflected, then further regular simplices are formed.

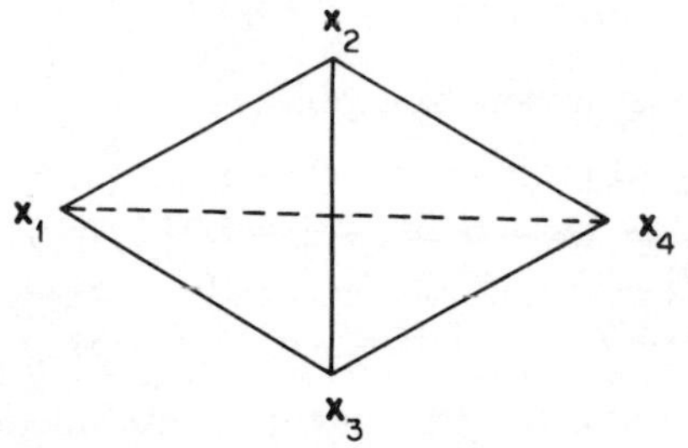

Figure 3.1 Reflection of vertex $\mathbf{x}_1$ in regular simplex

If $f(\mathbf{x})$ is a linear function of $\mathbf{x}$ in the neighbourhood of the initial simplex, then

$$f(\mathbf{x}_4) = f(\mathbf{x}_2 + \mathbf{x}_3 - \mathbf{x}_1) = f(\mathbf{x}_2) + f(\mathbf{x}_3) - f(\mathbf{x}_1). \tag{3.2}$$

Suppose that

$$f(\mathbf{x}_1) > \max\{f(\mathbf{x}_2), f(\mathbf{x}_3)\}.$$

Then equations (3.2) show that

$$f(\mathbf{x}_4) < f(\mathbf{x}_2) \quad \text{and} \quad f(\mathbf{x}_4) < f(\mathbf{x}_3).$$

Since $f(\mathbf{x})$ is to be minimized, it is advantageous to replace $\mathbf{x}_1$ by $\mathbf{x}_4$. However, when $f(\mathbf{x})$ is not linear in the neighbourhood of the current simplex, there is no guarantee that $f(\mathbf{x}_4) < f(\mathbf{x}_1)$. Nevertheless, the method of Spendley, Hext and Himsworth is based entirely on the above results for locally linear functions.

The recommended procedure for the minimization of $f(\mathbf{x})$ when $\mathbf{x} = [x_1, x_2]$ is therefore as follows:

1. Estimate the coordinates of the optimal point $\mathbf{x}^*$ and let this estimated point be one of the vertices of the initial simplex. Scale the variables if necessary and choose the remaining vertices of the initial simplex.
2. Evaluate $f(\mathbf{x})$ at the vertices of the initial simplex.
3. Replace the vertex with the highest function value by its reflection and evaluate $f(\mathbf{x})$ at the new vertex.
4. Repeat step 3 for each new simplex.
5. If the highest function value occurs at the new vertex, reflect the vertex with the second highest function value. This rule avoids oscillations of the type $\mathbf{x}_r \rightarrow \mathbf{x}_s \rightarrow \mathbf{x}_r \rightarrow \mathbf{x}_s \ldots$.
6. If one vertex persists for four iterations, reduce the size of the most recently formed simplex by halving the distances of the remaining vertices from that vertex. This step is called a *contraction*.
7. Stop after a prescribed number of contractions have been carried out.

The n-dimensional case

Let the vertices of a regular simplex in E^n be $\mathbf{x}_1, \ldots, \mathbf{x}_{n+1}$. The *reflection* of the vertex $\mathbf{x}_p$ is defined as

$$\mathbf{x}_q = 2\mathbf{x}_c - \mathbf{x}_p, \tag{3.3}$$

where

$$\mathbf{x}_c = \frac{1}{n}(\mathbf{x}_1 + ... + \mathbf{x}_{p-1} + \mathbf{x}_{p+1} + ... + \mathbf{x}_{n+1})$$

is the centroid of the remaining vertices. Equation (3.3) replaces equation (3.1) in rules 3 and 5 above. A *contraction* about a persistent vertex $\mathbf{x}_k$ consists in replacing the vertices $\mathbf{x}_p$ of a simplex by

$$\mathbf{x}_q = \tfrac{1}{2}(\mathbf{x}_p + \mathbf{x}_k).$$

That is, the vertex $\mathbf{x}_k$ remains fixed while the linear dimensions of the simplex are halved. Spendley, Hext and Himsworth suggest that a contraction be carried out whenever a vertex persists for more than m iterations, where m is the smallest integer exceeding $(1{\cdot}65n + 0{\cdot}05n^2)$; cf. step 6 above. This expression is an approximation for the maximum number of iterations for which a vertex is likely to persist; hence it is reasonable to assume that any 'older' vertex is in the vicinity of $\mathbf{x}^*$.

The method of Spendley, Hext and Himsworth is sometimes called the 'Simplex Method'—not to be confused with the simplex method for linear programming. The 'Complex Method', due to Box,[8] is a modification of Spendley, Hext and Himsworth's method which allows for constraints on the variables.

A major disadvantage of Spendley, Hext and Himsworth's method is that the use of regular simplices limits the movement of the current point in each iteration. For example, once a favourable direction has been found, it would be more efficient to move further in that direction than the regular simplex pattern allows. This improvement is made possible by using non-regular simplices, as in Nelder and Mead's method which is described in the next section.

3.5 NELDER AND MEAD'S METHOD

In 1965, Nelder and Mead[56] increased the efficiency of Spendley, Hext and Himsworth's method by allowing the simplices to become non-regular. Their method is one of the most efficient pattern search methods currently available and has been found to work particularly well if the number of variables does not exceed five or six.

Consider again the problem of minimizing $f(\mathbf{x})$. Let $\mathbf{x}_1$ be an initial estimate of $\mathbf{x}^*$ and let the vertices of the initial simplex be $\mathbf{x}_1, ..., \mathbf{x}_{n+1}$, where

$$\mathbf{x}_{j+1} = \mathbf{x}_1 + h_j\mathbf{e}_j \qquad (j = 1, ..., n),$$

the $\mathbf{e}_j$ are the usual unit coordinate vectors and the scalars h_j are chosen so as to equalize, as far as possible, the quantities

$$|f(\mathbf{x}_1 + h_j\mathbf{e}_j) - f(\mathbf{x}_1)|.$$

In the current simplex, let

$\mathbf{x}_h$ be the vertex with the highest function value,
$\mathbf{x}_s$ be the vertex with the second highest function value,
$\mathbf{x}_l$ be the vertex with the lowest function value,
$\mathbf{x}_c$ be the centroid of all the vertices except $\mathbf{x}_h$, i.e.

$$\mathbf{x}_c = \frac{1}{n} \sum_{\substack{j=1 \\ j \neq h}}^{n+1} \mathbf{x}_j.$$

Also, let $y = f(\mathbf{x})$, $y_h = f(\mathbf{x}_h)$, etc. Then the procedure recommended by Nelder and Mead for the minimization of $f(\mathbf{x})$ is as follows:

1. Choose the vertices of the initial simplex as described above and evaluate $f(\mathbf{x})$ at each vertex.

2. *Reflection.* Reflect $\mathbf{x}_h$ (Figure 3.2) using a reflection factor $\alpha > 0$, i.e. find $\mathbf{x}_0$ such that

$$\mathbf{x}_0 - \mathbf{x}_c = \alpha(\mathbf{x}_c - \mathbf{x}_h),$$

or

$$\mathbf{x}_0 = (1+\alpha)\mathbf{x}_c - \alpha\mathbf{x}_h \qquad (\alpha > 0).$$

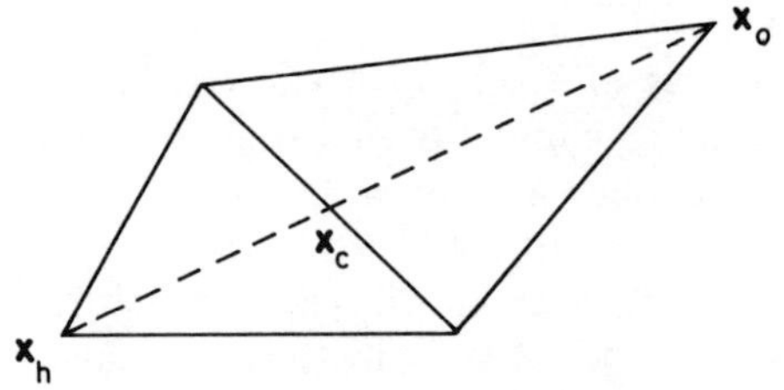

Figure 3.2 Reflection of vertex $\mathbf{x}_h$ in non-regular simplex

3. If $y_l \leqslant y_0 \leqslant y_s$, replace $\mathbf{x}_h$ by $\mathbf{x}_0$ and return to step 2.

4. *Expansion.* If $y_0 < y_l$, expand the simplex (Figure 3.3) using an expansion factor $\gamma > 1$, i.e. find $\mathbf{x}_{00}$ such that

$$\mathbf{x}_{00} - \mathbf{x}_c = \gamma(\mathbf{x}_0 - \mathbf{x}_c),$$

or

$$\mathbf{x}_{00} = \gamma\mathbf{x}_0 + (1-\gamma)\mathbf{x}_c \qquad (\gamma > 1).$$

(a) If $y_{00} < y_l$, replace $\mathbf{x}_h$ by $\mathbf{x}_{00}$ and return to step 2.

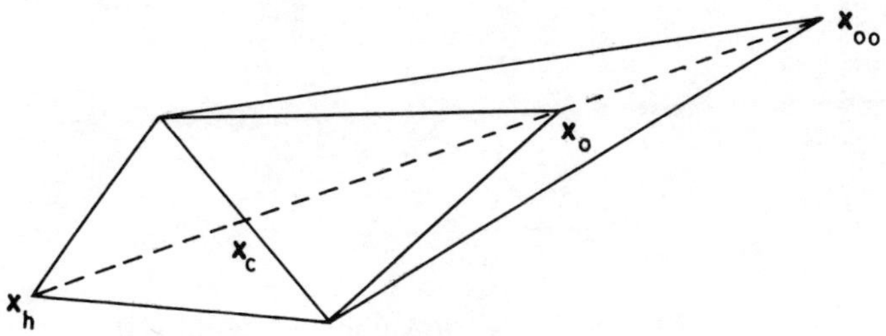

Figure 3.3 Expansion of non-regular simplex

(b) If $y_{00} \geqslant y_l$, replace $\mathbf{x}_h$ by $\mathbf{x}_0$ and return step 2.

5. *Contraction.* If $y_0 > y_s$, contract the simplex, using a contraction factor $\beta\,(0 < \beta < 1)$. There are two cases to consider:

(a) If $y_0 < y_h$ (Figure 3.4), find $\mathbf{x}_{00}$ such that

$$\mathbf{x}_{00} - \mathbf{x}_c = \beta(\mathbf{x}_0 - \mathbf{x}_c),$$

or

$$\mathbf{x}_{00} = \beta\mathbf{x}_0 + (1-\beta)\mathbf{x}_c \qquad (0 < \beta < 1).$$

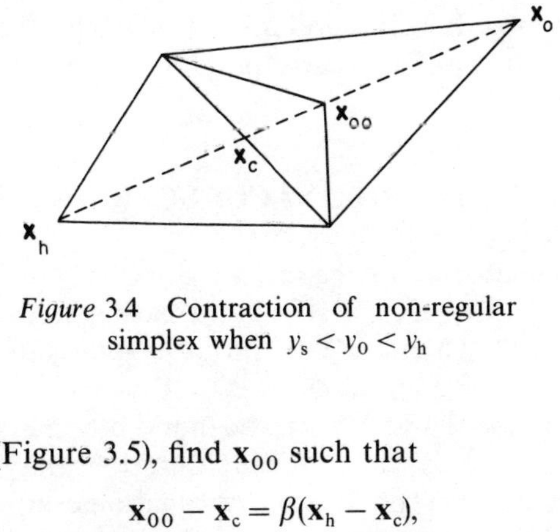

Figure 3.4 Contraction of non-regular simplex when $y_s < y_0 < y_h$

(b) If $y_0 \geqslant y_h$ (Figure 3.5), find $\mathbf{x}_{00}$ such that

$$\mathbf{x}_{00} - \mathbf{x}_c = \beta(\mathbf{x}_h - \mathbf{x}_c),$$

or

$$\mathbf{x}_{00} = \beta\mathbf{x}_h + (1-\beta)\mathbf{x}_c \qquad (0 < \beta < 1).$$

Whether 5(a) or 5(b) is used, there are again two cases to consider:

(c) If $y_{00} < y_h$ and $y_{00} < y_0$, replace $\mathbf{x}_h$ by $\mathbf{x}_{00}$ and return to step 2.

(d) If $y_{00} \geqslant y_h$ or $y_{00} > y_0$, reduce the size of the simplex by halving the distances from $\mathbf{x}_l$ and return to step 2.

Nelder and Mead suggest values of $\alpha = 1$, $\beta = \frac{1}{2}$ and $\gamma = 2$ for the reflection,

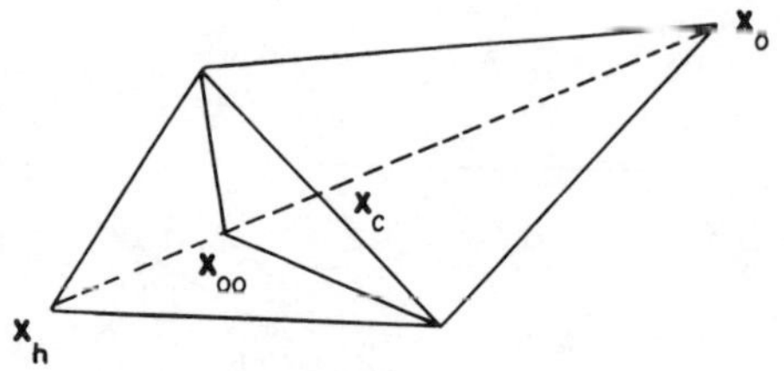

Figure 3.5 Contraction of non-regular simplex when $y_o \geqslant y_h$

contraction and expansion factors, respectively. A convenient convergence criterion is that the calculations terminate when the standard deviation of $y_1, \ldots, y_{n+1}$ is less than some prescribed value $\varepsilon > 0$, i.e. when

$$s = \sum_{k=1}^{n+1} \left\{ (y_k - \bar{y})^2 / n \right\}^{1/2} < \varepsilon,$$

where

$$\bar{y} = \left(\sum_{k=1}^{n+1} y_k \right) / (n+1).$$

Box, Davies and Swann[10] suggest a safer criterion: evaluate s after every K function evaluations, K being prescribed; stop when two successive values of s are each less than ε and the corresponding values of $\bar{y}$ differ by less than some prescribed amount.

3.6 FIBONACCI SEARCH

Suppose that a function $\phi(X)$ of a real variable X attains its maximum value in the interval $[0, L]$ at $X = X^*$, is strictly increasing for $X < X^*$ and is strictly decreasing for $X > X^*$. Then $\phi(X)$ is said to be *unimodal* in $[0, L]$. Note that $\phi(X)$ may or may not be continuous at $X = X^*$.

When $\phi(X)$ is unimodal and X^* is to be found by means of function evaluations and comparisons only, the following questions may be asked:

(a) For a given number of function evaluations, how should successive values of X be chosen so as to minimize the error in the estimate of X^*?

(b) Given an upper bound for the error in the estimate of X^*, how should successive values of X be chosen so as to minimize the number of function evaluations?

These questions are answered by considering the more general question:

(c) Given an upper bound for the error in the estimate of X^* and given the number of function evaluations, what is the greatest value of L for which this number of function evaluations is sufficient?

It was shown by Kiefer[50] that the answers to all these questions are related to the Fibonacci sequence $\{F_n\}$, defined by

$$F_0 = F_1 = 1, \quad F_n = F_{n-1} + F_{n-2} \qquad (n \geqslant 2). \tag{3.4}$$

This celebrated sequence was discovered by Leonardo of Pisa (1175–1230), otherwise known as Fibonacci, during an investigation of the rabbit population problem. The first sixteen terms of the sequence are

1, 1, 2, 3, 5, 8, 13, 21, 34, 55, 89, 144, 233, 377, 610, 987.

Solving the difference equation (3.4), we find

$$F_n = \frac{\sqrt{5}}{2^{n+1}.5}[(\sqrt{5}+1)^{n+1} + (-1)^n(\sqrt{5}-1)^{n+1}], \quad n=0,1,2,\ldots.,$$

and from this formula, or directly from (3.4), we find

$$\lim_{n\to\infty} \frac{F_n}{F_{n+1}} = \frac{\sqrt{5}-1}{2},$$

which is the well-known Golden Ratio (known to the Pythagoreans in 500 B.C.) in which the point C divides the line segment AB when AC/CB = AB/AC.

Theorem 3.1 provides the answer to question (c) in the case where $\phi(X)$ is a function of the continuous variable X; the answers to questions (a) and (b) then follow immediately. Theorem 3.2 gives the corresponding result for the case where $\phi(X)$ is defined on a discrete set of points.

***Theorem* 3.1**

Let $\phi(X)$ be a unimodal function defined on a closed interval $[0, L_n]$, where L_n is such that X^* can be located with an error of less than unity by making at most n evaluations of $\phi(X)$. Let

$$F_n = \text{l.u.b. } L_n.$$

Then the sequence $\{F_n\}$ is the Fibonacci sequence (3.4).

Proof

The proof is inductive. If $n = 0$ or 1, i.e. if we have no value of $\phi(X)$ or only one value of $\phi(X)$, then we have no information about X^* and hence $F_0 = F_1 = 1$ (the 1 being the given length of the final interval of uncertainty).

If $n \geqslant 2$, we evaluate $\phi(X)$ at $X = X_1$ and $X = X_2$, where $X_1, X_2 \in (0, L_n)$ and $X_1 < X_2$. These two values of X are to be determined subsequently.

If $n = 2$ and $L_2 = 2 - \varepsilon$, where $0 < \varepsilon < 1$, we can take $X_1 = 1 - \frac{1}{2}\varepsilon$ (the mid-

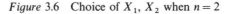

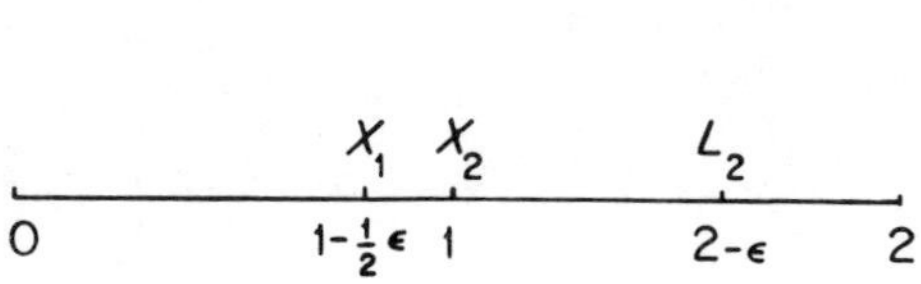

Figure 3.6 Choice of X_1, X_2 when $n = 2$

point of the interval) and $X_2 = 1$ (Figure 3.6). This choice of X_1 and X_2 locates X^* with an error of less than unity: in the interval $(0,1)$ if $\phi(X_1) > \phi(X_2)$ and in the interval $(1-\frac{1}{2}\varepsilon, 2-\varepsilon)$ if $\phi(X_1) < \phi(X_2)$. Hence

$$\text{l.u.b. } L_2 \geqslant 2.$$

On the other hand,

$$\text{l.u.b. } L_2 < 2 + \delta$$

for any $\delta > 0$, for if $L_2 > 2$ it is impossible to find two overlapping segments each of length unity which together cover the interval $[0, L_2]$. Therefore

$$F_2 = 2 = F_1 + F_0,$$

which begins the induction.

Now assume that

$$F_k = F_{k-1} + F_{k-2} \qquad (k=2,\ldots.,n-1). \tag{3.5}$$

We have to show that equation (3.5) is also true for $k = n$, and we do this by proving

(i) $F_n \leqslant F_{n-1} + F_{n-2}$,
(ii) $F_n \geqslant F_{n-1} + F_{n-2}$.

Suppose we evaluate $\phi(X)$ at $X = X_1$ and $X = X_2$, where $X_1, X_2 \in (0, L_n)$ and $X_1 < X_2$. If $\phi(X_1) > \phi(X_2)$ then the remaining interval of uncertainty is $[0, X_2)$, and we already know $\phi(X_1)$, where $X_1 \in (0, X_2)$. Hence $X_1 < F_{n-2}$, since X^* may lie in the interval $[0, X_1]$ and we are allowed only $(n-2)$ further evaluations. Also $X_2 < F_{n-1}$, since X^* must be located with an error of less than unity in $[0, X_2]$ by evaluating $\phi(X)$ at $X = X_1$ and at $(n-2)$ other points at most.

Similarly, if $\phi(X_1) < \phi(X_2)$, then the remaining interval of uncertainty is $(X_1, L_n]$, and

$$L_n - X_1 < F_{n-1}.$$

Hence

$$\begin{aligned} L_n &< F_{n-1} + X_1 \\ &< F_{n-1} + F_{n-2}. \end{aligned}$$

It follows that

$$F_n = \text{l.u.b. } L_n \leqslant F_{n-1} + F_{n-2}, \tag{3.6}$$

and (i) is proved.

To prove (ii), we can choose

$$L_n = (1 - \tfrac{1}{2}\varepsilon)(F_{n-1} + F_{n-2}), \tag{3.7}$$

with

$$X_1 = (1 - \tfrac{1}{2}\varepsilon)F_{n-2}, \quad X_2 = (1 - \tfrac{1}{2}\varepsilon)F_{n-1}, \tag{3.8}$$

where $0 < \varepsilon < 1$. The choice of the factor $(1 - \frac{1}{2}\varepsilon)$ in equation (3.7) is consistent with $L_2 = 2 - \varepsilon$, as used previously. Equation (3.7) shows that if L_n is taken to be any fixed number less than $(F_{n-1} + F_{n-2})$, then it can always be increased, since ε may be made arbitrarily small. It follows that

$$F_n = \text{l.u.b. } L_n \geqslant F_{n-1} + F_{n-2}. \tag{3.9}$$

The inequalities (3.6) and (3.9) give

$$F_n = F_{n-1} + F_{n-2}. \tag{3.10}$$

Thus equation (3.5) is true for $k = n$, and hence for all $k \geqslant 2$. But $F_0 = F_1 = 1$, and so $\{F_n\}$ is the Fibonacci sequence (3.4).

The following method for maximizing $f(\mathbf{x})$ on the interval $0 \leqslant x \leqslant l$ is suggested by the proof of Theorem 3.1. For any ε, $0 < \varepsilon \ll 1$, suppose that

$$\phi(X) \equiv f\left(\frac{lX}{L_n}\right)$$

is defined on the interval $0 \leqslant X \leqslant L_n$, where

$$L_n = (1 - \tfrac{1}{2}\varepsilon)F_n.$$

This formula comes from equations (3.7) and (3.10). Using X_1 and X_2 from equation (3.8), compare the values $\phi(X_1)$ and $\phi(X_2)$, leaving an interval of uncertainty of length

$$L_{n-1} = (1 - \tfrac{1}{2}\varepsilon)F_{n-1},$$

together with the value of $\phi(X)$ at one of the two optimal points of evaluation for this smaller interval. One further evaluation and comparison leaves an interval of uncertainty of length

$$L_{n-2} = (1 - \tfrac{1}{2}\varepsilon)F_{n-2};$$

proceeding in this way we have, after $(n - k + 1)$ evaluations,

$$L_k = (1 - \tfrac{1}{2}\varepsilon)F_k \qquad (2 \leqslant k \leqslant n).$$

In particular, after $(n - 1)$ evaluations, we have

$$L_2 = (1 - \tfrac{1}{2}\varepsilon)F_2 = 2 - \varepsilon.$$

The nth and final evaluation is at one of the points X_1, X_2 indicated in Figure 3.6; and hence the final interval of uncertainty is less than unity. From Figure 3.6 we see that the final error in X^* is not greater than $\max\{1 - \frac{1}{2}\varepsilon, \frac{1}{2}\varepsilon\}$.

Normally, it is sufficient to locate X^* with an error of *not more than* unity. This corresponds to setting $\varepsilon = 0$ in equations (3.7). The points X_1 and X_2 in Figure 3.6 now coincide, and so the number of function evaluations is reduced by one, to $(n - 1)$, for a function defined on an interval of length F_n. The following example illustrates this case.

Example 3.2

Maximize $f(x) \equiv e^x - 2x^2$ *over the interval* $[0,1]$ *with an error in* x^* *of not more than* 0·05.

Solution

A sketch-graph shows that $f(x)$ is unimodal over the interval $[0,1]$. In applying Theorem 3.1, we first scale the variable x so that the unit of length in the Fibonacci search is the allowable error 0·05. Thus the initial interval of uncertainty is of length $L = 1/0{\cdot}05 = 20$. This length must be increased to $L_n = F_n = 21$, the next higher Fibonacci number, whence $n = 7$, and six function evaluations are required. [When the initial interval of uncertainty is extended in this way, it is assumed that $\phi(X) = \mp M$ in the extended part of the interval for a maximizing or minimizing problem, respectively, where M is a number greater than any value of $|\phi(X)|$ in the initial interval.]

The calculations are set out in Table 3.1, in which L_n is the length of the interval of uncertainty at the beginning of the current step, D_1 and D_2 are its end-points ($L_n = D_2 - D_1$), and X_1, X_2 are the points at which $\phi(X)$ is evaluated in each step. Using equations (3.7) and (3.8) with $\varepsilon = 0$, and noting that the left-hand end of the interval of uncertainty is at $X = D_1$, we have

$$L_n = F_n, \quad X_1 = D_1 + F_{n-2}, \quad X_2 = D_1 + F_{n-1}.$$

Also,

$$x = 0{\cdot}05X, \quad \phi(X) \equiv f(0{\cdot}05X) = f(x).$$

For each step, the greater of the two values of $f(x)$ is underlined.

Table 3.1 Solution of Example 3.2

n	L_n	D_1	D_2	X_1	X_2	x_1	x_2	$f(x_1)$	$f(x_2)$
7	21	0	21	8	13	0·4	0·65	<u>1·171825</u>	1·070541
6	13	0	13	5	8	0·25	0·4	1·159025	<u>1·171825</u>
5	8	5	13	8	10	0·4	0·5	<u>1·171825</u>	1·148721
4	5	5	10	7	8	0·35	0·4	<u>1·174068</u>	1·171825
3	3	5	8	6	7	0·3	0·35	1·169859	<u>1·174068</u>

Table 3.1 shows that $x^* = 0{\cdot}35$, $f(x^*) = 1{\cdot}174068$, with an error in x^* of not more than 0·05. In fact, $x^* = 0{\cdot}357403$, $f(x^*) = 1{\cdot}174138$, each correct to six places of decimals.

We now turn to the case where the function to be maximized is defined on a discrete set of points. The proof that Fibonacci search is optimal for this case is provided by the following theorem.

***Theorem* 3.2**

Let $\phi(X)$ be a unimodal function defined on a discrete set of H_n points, where H_n is such that X^* can be determined by making at most n evaluations of $\phi(X)$. If $M_n = \max H_n$, then

$$M_n = F_{n+1} - 1 \qquad (n \geqslant 1),$$

where $\{F_n\}$ is the Fibonacci sequence (3.4).

Proof

Denote the discrete points at which $\phi(X)$ is defined by the H_n integers X_r $(r = 1, ..., H_n)$, not necessarily in numerical order. Obviously,

$$M_1 = 1 = F_2 - 1, \quad M_2 = 2 = F_3 - 1.$$

Also (Figure 3.7),

$$M_3 = 4 = F_4 - 1,$$

for if $n = 3$ we evaluate $\phi(X_1)$ and $\phi(X_2)$, then if $\phi(X_1) > \phi(X_2)$ we discard X_4 and evaluate $\phi(X_3)$; on the other hand, if $\phi(X_1) < \phi(X_2)$ we discard X_3 and evaluate $\phi(X_4)$.

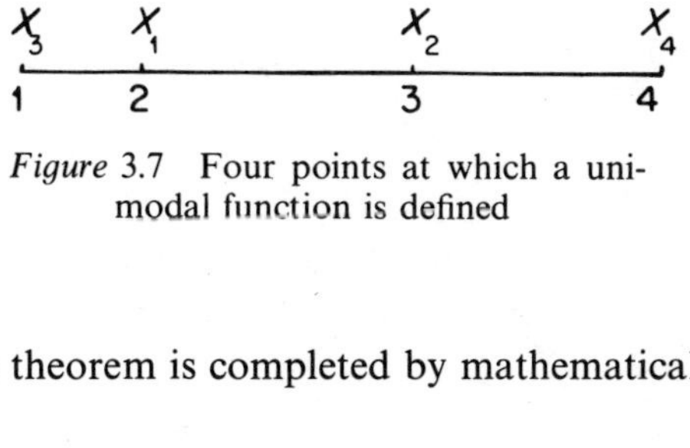

Figure 3.7 Four points at which a unimodal function is defined

The proof of the theorem is completed by mathematical induction. Assume that

$$M_k = F_{k+1} - 1 \qquad (k = 1, ..., n-1).$$

Suppose that the interval of uncertainty is AD (Figure 3.8) and that we evaluate $\phi(X_1)$ and $\phi(X_2)$. After comparing the values $\phi(X_1)$ and $\phi(X_2)$, the remaining

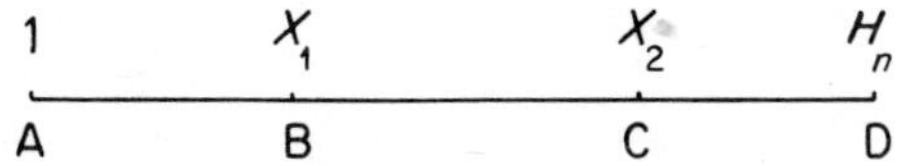

Figure 3.8 Choice of X_1, X_2 in the interval $[1, H_n]$

interval is either AC or BD, and hence

$$X_2 - 1 = H_n - X_1 \leqslant M_{n-1} = F_n - 1, \tag{3.11}$$

where we have used the induction hypothesis. Similarly, at the end of the next step, the interval of uncertainty is reduced to AB or CD or some equivalent interval, and hence

$$X_1 - 1 = H_n - X_2 \leqslant M_{n-2} = F_{n-1} - 1. \tag{3.12}$$

From (3.11) and (3.12), we obtain

$$H_n \leqslant X_2 + F_{n-1} - 1 \leqslant F_n + F_{n-1} - 1 = F_{n+1} - 1. \tag{3.13}$$

Replacing the inequalities in (3.11) and (3.12) by equations, we see that the maximum value of H_n is attained when

$$X_1 = F_{n-1}, \quad X_2 = F_n,$$

and then (3.13) gives

$$M_n = \max H_n = F_{n+1} - 1,$$

which completes the induction.

Example 3.3

Since $F_7 = 21$, the maximum value of a unimodal function defined on twenty discrete points can be determined by making six evaluations:

$$M_6 = F_7 - 1 = 20.$$

A function defined on twenty-one discrete points requires seven evaluations.

Example 3.4

Find the maximum value of the function

$$\phi(X) \equiv X^3 - 3456\,X^2 + 2{,}345{,}678\,X$$

defined on the set of points $X = 1,2,\ldots,1500$.

Table 3.2 Solution of Example 3.4

n	F_{n+1}	H_n	D_1	D_2	X_1	X_2	$\phi(X_1)$	$\phi(X_2)$
15	1597	1596	1	1596	610	987	371,866,980	−90,039,075
14	987	986	1	986	377	610	446,705,415	371,866,980
13	610	609	1	609	233	377	371,569,527	446,705,415
12	377	376	234	609	377	466	446,705,415	443,789,508
11	233	232	234	465	322	377	430,362,660	446,705,415
10	144	143	323	465	377	411	446,705,415	449,709,213
9	89	88	378	465	411	432	449,709,213	448,981,920
8	55	54	378	431	398	411	449,180,412	449,709,213
7	34	33	399	431	411	419	449,709,213	449,660,325
6	21	20	399	418	406	411	449,595,468	449,709,213
5	13	12	407	418	411	414	449,709,213	449,724,060
4	8	7	412	418	414	416	449,724,060	449,711,808
3	5	4	412	415	413	414	449,723,547	449,724,060
2	3	2	414	415	414	415	449,724,060	449,720,145

Solution

The function is unimodal. Also, $F_{16} = 1597$ and hence fifteen evaluations are required:

$$M_{15} = F_{16} - 1.$$

In Table 3.2, $H_n = F_{n+1} - 1$ is the number of points in the current interval of uncertainty, D_1 and D_2 are the first and last points in this interval ($H_n = D_2 - D_1 + 1$) and X_1, X_2 are the points of evaluation. Note that

$$X_2 - X_1 = F_{n-2}, \quad X_1 + X_2 = D_1 + D_2.$$

Table 3.2 shows that

$$X^* = 414, \quad \phi(X^*) = 449{,}724{,}060.$$

Example 3.5

Show how to use the result of Example 3.4 *to find the maximum value of the function*

$$f(x) \equiv x^3 - 34{\cdot}56x^2 + 234{\cdot}5678x$$

defined on the set of points $x = 0{\cdot}01, 0{\cdot}02, \ldots\ldots 15{\cdot}00$.

Finally, it should be noted that, although we have referred throughout this section to a function of one variable, the results can be applied slightly more generally to the problem of maximizing a function $f(\mathbf{x})$ along the line $\mathbf{x} = \mathbf{x}_k + \lambda\mathbf{d}$, where $\mathbf{x}_k$ is the current point and $\mathbf{d}$ is a given direction of search. This is the typical *linear search* or *one-dimensional search*. Similar remarks apply to any optimization technique for a function of one variable.

3.7 GOLDEN SECTION SEARCH

Golden Section search is based on the fact that

$$r = \lim_{n\to\infty} \frac{F_n}{F_{n+1}} = \frac{\sqrt{5}-1}{2} \doteqdot 0{\cdot}618034,$$

where F_n and F_{n+1} are successive terms of the Fibonacci sequence (3.4). In this search technique, the interval of uncertainty is reduced by the constant factor r at each step; thus Golden Section search is a limiting form of Fibonacci search. In general, it can be used only on functions of a continuous variable. Note that r satisfies

$$r^2 + r - 1 = 0.$$

Let $\phi(X)$ be a unimodal function of a continuous variable X defined on the closed interval $[0, L_n]$. The points of evaluation in Golden Section search may be found by dividing equations (3.8) by equation (3.7) and letting $n \to \infty$. This gives

$$X_1 = r^2 L_n, \quad X_2 = r L_n. \tag{3.14}$$

The Golden Section search to maximize the unimodal function $\phi(X)$ on the interval $[0, L_n]$ therefore proceeds as follows:

1. Evaluate $\phi(X_1)$ and $\phi(X_2)$, where X_1 and X_2 are given by equations (3.14).
2. If $\phi(X_1) > \phi(X_2)$, discard the interval $(X_2, L_n]$. The remaining interval is of length rL_n and

$$X_1 = r(rL_n) = r^2 L_n$$

is one of the points of evaluation on it. The other point of evaluation is

$$X = r^2(rL_n) = r^3 L_n.$$

On the other hand, if $\phi(X_2) > \phi(X_1)$, discard the interval $[0, X_1)$. The remaining interval is again of length rL_n and X_2 is one of the points of evaluation on it, since

$$L_n - X_1 = (1 - r^2)L_n = rL_n$$

and

$$X_2 - X_1 = rL_n - r^2 L_n = r^3 L_n = r^2(rL_n).$$

The other point of evaluation is

$$X = X_1 + r(rL_n) = 2r^2 L_n.$$

3. Repeat steps 1 and 2 for the successive remaining intervals of lengths $rL_n, r^2 L_n, r^3 L_n, \ldots,$ until the desired accuracy in X^* is attained.

It is evident that Golden Section search is simpler, though less efficient, than Fibonacci search. In particular, when optimizing the function $f(x)$ there is no need to scale the variable x.

Example 3.6

Find the minimum value of $f(x) \equiv x^2 + 2e^{-x}$, *with an error of not more than* 0·02 *in* x^*.

Solution

A sketch-graph shows that $f(x)$ is unimodal over the interval $-\infty < x < \infty$; also,

$$f(0) = 2, \ f(1) = 1 + \frac{2}{e}, \ f(2) = 4 + \frac{2}{e^2}.$$

We therefore take the initial interval of uncertainty to be $[0, 2]$; this must be reduced finally to an interval of length 0·02. Thus n steps are required, where n is the smallest integer satisfying $2(0{\cdot}618034)^n \leqslant 0{\cdot}02$, giving $n = 10$. In Table 3.3,

L_n is the length of the interval of uncertainty at the beginning of the current step, d_1, d_2 are its end-points ($L_n = d_2 - d_1$) and x_1, x_2 are the points of evaluation at each step. Note that

$$x_1 + x_2 = d_1 + d_2.$$

Table 3.3 Solution of Example 3.6

n	L_n	d_1	d_2	x_1	x_2	$f(x_1)$	$f(x_2)$
10	2	0	2	0·763932	1·236068	1·515254	2·108913
9	1·236068	0	1·236068	0·472136	0·763932	1·470250	1·515254
8	0·763932	0	0·763932	0·291796	0·472136	1·578987	1·470250
7	0·472136	0·291796	0·763932	0·472136	0·583592	1·470250	1·456361
6	0·291796	0·472136	0·763932	0·583592	0·652476	1·456361	1·467234
5	0·180340	0·472136	0·652476	0·541020	0·583592	1·457011	1·456361
4	0·111456	0·541020	0·652476	0·583592	0·609903	1·456361	1·458789
3	0·068884	0·541020	0·609903	0·567331	0·583592	1·455938	1·456361
2	0·042572	0·541020	0·583592	0·557281	0·567331	1·456091	1·455938
1	0·026311	0·557281	0·583592	0·567331	0·573542	1·455938	1·456002

Table 3.3 shows that x^* lies in the interval [0·557281, 0·573542] after the step $n = 1$. The best value for $f(x^*)$ to this accuracy is

$$f(x_1) = f(0{\cdot}567331) = 1{\cdot}455938.$$

To six places of decimals, the correct optimal values are

$$x^* = 0{\cdot}567143, \quad f(x^*) = 1{\cdot}455938.$$

Thus x^* and $f(x^*)$ have been found correct to three and six places of decimals, respectively.

3.8 POWELL'S QUADRATIC INTERPOLATION METHOD

Consider the linear search problem of minimizing the function $f(\mathbf{x})$ along the line $\mathbf{x} = \mathbf{x}_k + \lambda\mathbf{d}$, where $\mathbf{x}_k$ is the current point and $\mathbf{d}$ is a given search direction. In 1964, Powell[62] published a simple algorithm for determining the minimizing value of λ, using quadratic interpolation together with a few common-sense rules. This algorithm forms part of Powell's more general method for finding the minimum value of a function $f(\mathbf{x})$ without calculating derivatives (see Section 4.9). However, it may also be used in conjunction with any gradient method or, more generally, with any optimization technique that requires a one-dimensional search.

In each iteration, Powell's algorithm finds a quadratic function $y(\lambda)$ which takes the same values as $f(\mathbf{x}_k + \lambda\mathbf{d})$ for three current values of λ. Having found the value of λ ($\lambda = \lambda_m$, say) which minimizes $y(\lambda)$, one of the three current values of λ is discarded and is replaced by λ_m. The iterations continue until the desired accuracy is attained.

Let

$$f_\alpha = f(\mathbf{x}_k + \alpha\mathbf{d}),$$
$$f_\beta = f(\mathbf{x}_k + \beta\mathbf{d}),$$
$$f_\gamma = f(\mathbf{x}_k + \gamma\mathbf{d})$$

be function values at three points, not necessarily consecutive, on the line $\mathbf{x} = \mathbf{x}_k + \lambda\mathbf{d}$. Assume that the quadratic function

$$y(\lambda) \equiv y_0 + y_1\lambda + y_2\lambda^2 \tag{3.15}$$

takes the values $f_\alpha, f_\beta, f_\gamma$ at $\lambda = \alpha, \beta, \gamma$, respectively, i.e. assume

$$\begin{aligned} y_0 + y_1\alpha + y_2\alpha^2 &= f_\alpha, \\ y_0 + y_1\beta + y_2\beta^2 &= f_\beta, \\ y_0 + y_1\gamma + y_2\gamma^2 &= f_\gamma. \end{aligned} \tag{3.16}$$

Equations (3.16) give

$$\left.\begin{aligned} y_0 &= [\beta\gamma(\gamma-\beta)f_\alpha + \gamma\alpha(\alpha-\gamma)f_\beta + \alpha\beta(\beta-\alpha)f_\gamma]/\boldsymbol{\Delta}, \\ y_1 &= [(\beta^2-\gamma^2)f_\alpha + (\gamma^2-\alpha^2)f_\beta + (\alpha^2-\beta^2)f_\gamma]/\boldsymbol{\Delta}, \\ y_2 &= [(\gamma-\beta)f_\alpha + (\alpha-\gamma)f_\beta + (\beta-\alpha)f_\gamma]/\boldsymbol{\Delta}, \end{aligned}\right\} \tag{3.17}$$

where

$$\boldsymbol{\Delta} = (\alpha-\beta)(\beta-\gamma)(\gamma-\alpha).$$

The quadratic function $y(\lambda)$ of (3.15) has a turning point at $\lambda = -y_1/2y_2$ and has a minimum there if $y_2 > 0$. Hence, using equations (3.17), we find that this turning point is at $\lambda = \lambda_m$, where

$$\lambda_m = \frac{1}{2}\left[\frac{(\beta^2-\gamma^2)f_\alpha + (\gamma^2-\alpha^2)f_\beta + (\alpha^2-\beta^2)f_\gamma}{(\beta-\gamma)f_\alpha + (\gamma-\alpha)f_\beta + (\alpha-\beta)f_\gamma}\right], \tag{3.18}$$

and $y(\lambda)$ has a minimum there if

$$\frac{(\beta-\gamma)f_\alpha + (\gamma-\alpha)f_\beta + (\alpha-\beta)f_\gamma}{(\alpha-\beta)(\beta-\gamma)(\gamma-\alpha)} < 0. \tag{3.19}$$

Given an initial point $\mathbf{x}_k$ and a direction of search $\mathbf{d}$, Powell's quadratic interpolation method for minimizing the general function $f(\mathbf{x})$ on the line $\mathbf{x} = \mathbf{x}_k + \lambda\mathbf{d}$ is as follows:

1. Choose a step length $h|\mathbf{d}|$; the vector $\mathbf{d}$ need not be a unit vector.
2. Evaluate $f(\mathbf{x}_k)$ and $f(\mathbf{x}_k + h\mathbf{d})$.
3. If $f(\mathbf{x}_k) < f(\mathbf{x}_k + h\mathbf{d})$, evaluate $f(\mathbf{x}_k - h\mathbf{d})$. Otherwise, evaluate $f(\mathbf{x}_k + 2h\mathbf{d})$. Values of $f(\mathbf{x})$ are now known at three points on the line $\mathbf{x} = \mathbf{x}_k + \lambda\mathbf{d}$.
4. Find the turning point $\lambda = \lambda_m$ of the quadratic function $y(\lambda)$ fitted through these three points, using (3.18), and test for a minimum, using (3.19). Go to rule 5, 6 or 7, as appropriate.
5. If the point $\lambda = \lambda_m$ corresponds to a maximum of $y(\lambda)$ or if it corresponds

to a minimum which is at a greater distance than $H|\mathbf{d}|$ (where H is prescribed) from the nearest of the three current points, proceed as follows. Discard the point which is furthest from the turning point and obtain a new current point and function value by taking a step $H|\mathbf{d}|$ in the direction in which the function decreases; this step (Figure 3.9) is taken from the point furthest from (nearest to) the turning point when the turning point corresponds to a maximum (minimum). Return to rule 4.

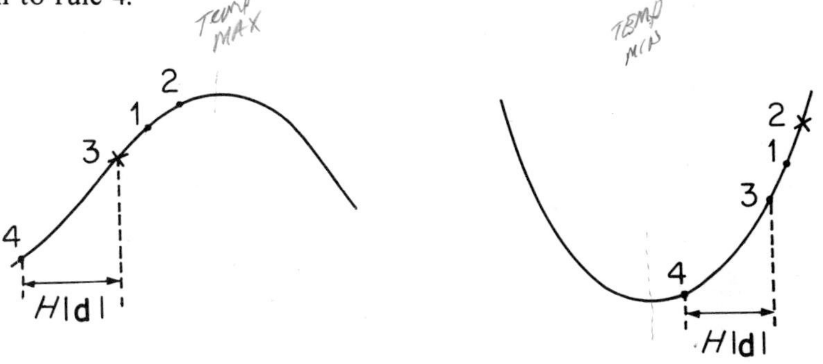

Figure 3.9 Rule 5 of Powell's quadratic interpolation method

6. If the point $\lambda = \lambda_m$ corresponds to a minimum of $y(\lambda)$ and if it is within a small prescribed distance $\varepsilon|\mathbf{d}|$ of the nearest point $\lambda = \alpha$, say, of the three current points, take

$$\min\{f(\mathbf{x}_k + \lambda_m \mathbf{d}), f(\mathbf{x}_k + \alpha\mathbf{d})\}$$

as the required minimum value of $f(\mathbf{x})$.

7. If the point $\lambda = \lambda_m$ corresponds to a minimum of $y(\lambda)$ to which neither rule 5 nor rule 6 applies, i.e. if it is not further than $H|\mathbf{d}|$ from the nearest of the three current points, but not within $\varepsilon|\mathbf{d}|$ of it, discard the point with the highest function value and replace it by $\lambda = \lambda_m$. Return to rule 4.

Note. It is always desirable to locate the next turning point by interpolation rather than by extrapolation; an exception is made to rule 7 to allow for this. In Figure 3.10, point 3 has the highest function value, though point 2 is discarded because the minimum lies between points 1 and 3.

As a very useful special case, the three current points on the line $\mathbf{x} = \mathbf{x}_k + \lambda\mathbf{d}$ may be equally spaced. It is then possible by suitable changes of origin and scale to take $\alpha = -1$, $\beta = 0$, $\gamma = 1$, and equation (3.18) reduces to

$$\lambda_m = \frac{f_\alpha - f_\gamma}{2(f_\alpha - 2f_\beta + f_\gamma)}. \tag{3.20}$$

From equations (3.15), (3.17) and (3.20), we find the turning value of $y(\lambda)$ in this case to be

Figure 3.10 Exception to rule 7 of Powell's quadratic interpolation method

$$y(\lambda_m) = f_\beta - \frac{(f_\alpha - f_\gamma)^2}{8(f_\alpha - 2f_\beta + f_\gamma)}, \tag{3.21}$$

and the condition (3.19) that it is a minimum reduces to

$$f_\alpha - 2f_\beta + f_\gamma > 0.$$

In applying Powell's quadratic interpolation method, the number of function evaluations may be reduced when more than one search has to be made in the same direction by noting that three function values are sufficient to predict the second derivative

$$\frac{\partial^2}{\partial \lambda^2}[f(\mathbf{x}_k + \lambda \mathbf{d})].$$

The prediction is

$$D = 2y_2 = \frac{2[(\gamma-\beta)f_\alpha + (\alpha-\gamma)f_\beta + (\beta-\alpha)f_\gamma]}{(\alpha-\beta)(\beta-\gamma)(\gamma-\alpha)}.$$

Then, if f_α, f_β and D are known, the minimum of $y(\lambda)$ is predicted to be at $\lambda = \lambda_m$, where

$$\lambda_m = \frac{1}{2}(\alpha+\beta) - \frac{(f_\alpha - f_\beta)}{D(\alpha-\beta)}. \tag{3.22}$$

Equation (3.22), which replaces equation (3.18) for the second and subsequent searches in the direction **d**, is derived by subtracting the second of equations (3.16) from the first and using the relations

$$\lambda_m = -y_1/2y_2 = -y_1/D.$$

Example 3.7

Use Powell's quadratic interpolation method to minimize

$$f(\mathbf{x}) \equiv x_1{}^4 - x_1{}^3 x_2 - x_1{}^2 x_2 x_3 + x_1 x_2 x_3 x_4$$

along the line $\mathbf{x} = \mathbf{a} + \lambda\mathbf{d}$, *where* $\mathbf{a} = [0, -1, -2, -3]$, $\mathbf{d} = [1, 2, 3, 4]$. *Take the current point to be* $\mathbf{x}_k = \mathbf{a}$; *also, take* $h = 0{\cdot}5$, $H = 2{\cdot}0$, $\varepsilon = 0{\cdot}001$.

Solution

First, we evaluate

$$f_\alpha = f(\mathbf{x}_k) = f(0, -1, -2, -3) = 0,$$
$$f_\beta = f(\mathbf{x}_k + h\mathbf{d}) = f(0{\cdot}5, 0, -0{\cdot}5, -1) = 0{\cdot}0625,$$
$$f_\gamma = f(\mathbf{x}_k - h\mathbf{d}) = f(-0{\cdot}5, -2, -3{\cdot}5, -5) = 15{\cdot}4375;$$

these are the function values for the first iteration. It is convenient to tabulate the subsequent results (Table 3.4). Every value of λ_m corresponds to a minimum of $y(\lambda)$, as may be verified from condition (3.19).

Table 3.4 Solution of Example 3.7

Iteration Number	α	β	γ	f_α	f_β	f_γ	λ_m
1	0	0·5	−0·5	0	0·0625	15·4375	0·2480
2	0	0·5	0·2480	0	0·0625	−0·3427	0·2395
3	0	0·2395	0·2480	0	−0·3543	−0·3427	0·1841
4	0	0·2395	0·1841	0	−0·3543	−0·4071	0·1758
5	0	0·1758	0·1841	0	−0·4106	−0·4071	0·1659
6	0	0·1758	0·1659	0	−0·4106	−0·4129	0·1632
7	0	0·1632	0·1659	0	−0·4132	−0·4129	0·1613
8	0	0·1632	0·1613	0	−0·4132	−0·4133	0·1607

The computation ends after iteration 8, since at this stage, $|\lambda_m - \gamma| = 0{\cdot}0006 < \varepsilon$. (The fact that $\alpha = 0$ throughout is fortuitous.) We conclude that the required minimum value of $f(\mathbf{x})$ to this accuracy is

$$f(\mathbf{x}^*) = \min\{f(\mathbf{x}_k + \lambda_m\mathbf{d}), f(\mathbf{x}_k + \gamma\mathbf{d})\} = -0{\cdot}4133,$$

the two function values in the final comparison agreeing to four places of decimals. Taking $\lambda^* = \frac{1}{2}(\lambda_m + \gamma)$, we find

$$\begin{aligned}\mathbf{x}^* = \mathbf{x}_k + \lambda^*\mathbf{d} &= [0, -1, -2, -3] + 0{\cdot}1610[1, 2, 3, 4]\\ &= [0{\cdot}1610, -0{\cdot}6780, -1{\cdot}5170, -2{\cdot}3560].\end{aligned}$$

Note that, in this example,

$$f(-\mathbf{x}) = f(\mathbf{x}),$$

and so there is another optimal point, symmetrical about the origin with respect to the first.

3.9 DAVIDON'S CUBIC INTERPOLATION METHOD

Powell's quadratic interpolation method, described in the previous section, uses values of $f(\mathbf{x})$ at three points on the line $\mathbf{x} = \mathbf{x}_k + \lambda\mathbf{d}$. In 1959, Davidon[24] devised a cubic interpolation method which uses function values at two points on this line, together with values of the directional derivatives of $f(\mathbf{x})$ along the line at the same two points. Thus Davidon's formulae are more elaborate than those of Powell; to compensate for this, Davidon's method usually finds the minimum of $f(\mathbf{x})$ in fewer iterations.

If the directional derivatives of $f(\mathbf{x})$ are relatively easy to evaluate, then Davidon's cubic interpolation method is generally to be preferred to Powell's quadratic interpolation method and is particularly recommended for use with the Davidon–Fletcher–Powell and Fletcher–Reeves methods of unconstrained optimization—see Sections 4.4 and 4.7.

Consider again the linear search problem of minimizing $f(\mathbf{x})$ along the line $\mathbf{x} = \mathbf{x}_k + \lambda\mathbf{d}$, where $\mathbf{x}_k$ is the current point and $\mathbf{d}$ is a given search direction. Suppose that $f_0 = f(\mathbf{x}_k)$ and $f_\alpha = f(\mathbf{x}_k + \alpha\mathbf{d})$ are known, where α is a given value of λ. Suppose also that the directional derivatives

$$G_0 = [\mathbf{g}(\mathbf{x}_k)]'\mathbf{d}, \qquad G_\alpha = [\mathbf{g}(\mathbf{x}_k + \alpha\mathbf{d})]'\mathbf{d},$$

where $\mathbf{g}(\mathbf{x}) = \nabla f(\mathbf{x})$, are known and that $G_0 < 0$.

The calculations for this method are in three distinct stages:

(a) The order of magnitude of λ_m, the minimizing value of λ, is estimated.

(b) Upper and lower bounds are found for λ_m.

(c) Cubic interpolation is used to find more precise bounds on λ_m.

With regard to (a), it is usual to take the parameter α as an initial approximation to λ_m, where

$$\alpha = \min\left\{\kappa, -\frac{2(f_0 - f_e)}{G_0}\right\}, \tag{3.23}$$

where κ is some representative magnitude for the problem (usually, $\kappa = 2$) and f_e is a preliminary estimate, preferably low rather than high, of $f(\mathbf{x}_k + \lambda_m\mathbf{d})$. The expression $-2(f_0 - f_e)/G_0$ in equation (3.23) is equal to λ_m in the ideal case where $f(\mathbf{x})$ is a quadratic function of $\mathbf{x}$ and where f_e is an exact estimate of the minimum value of $f(\mathbf{x})$ on the line $\mathbf{x} = \mathbf{x}_k + \lambda\mathbf{d}$.

To implement stages (b) and (c), Davidon approximates the function $f(\mathbf{x}_k + \lambda\mathbf{d})$ by a cubic $y(\lambda)$ which satisfies

$$y(0) = f_0, \;\; y(\alpha) = f_\alpha, \;\; y'(0) = G_0, \;\; y'(\alpha) = G_\alpha. \tag{3.24}$$

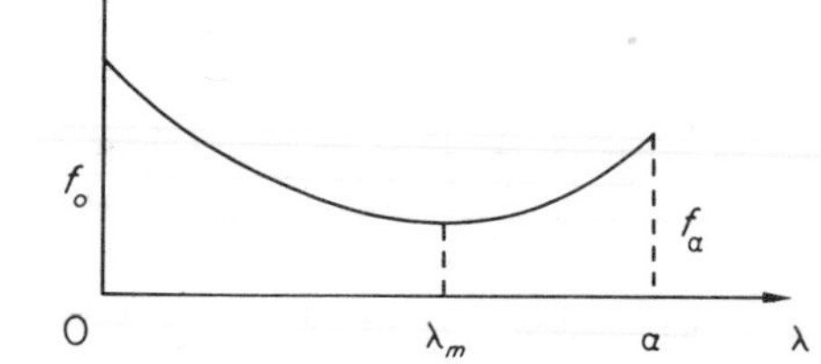

Figure 3.11 First form for $y(\lambda)$

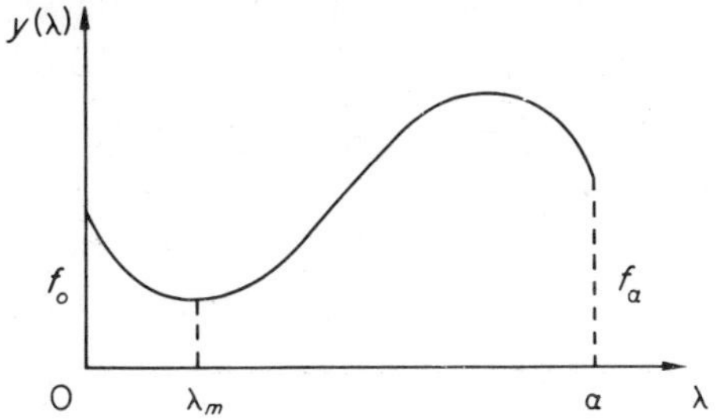

Figure 3.12 Second form for $y(\lambda)$

Since $G_0 < 0$, there is certainly a minimum of $y(\lambda)$ between $\lambda = 0$ and $\lambda = \alpha$ if $G_\alpha > 0$ (Figure 3.11) or if $f_\alpha > f_0$ (Figure 3.12). When neither of these conditions holds, the point $\mathbf{x}_k + \alpha\mathbf{d}$ is replaced by the point $\mathbf{x}_k + 2\alpha\mathbf{d}$. Repeating this procedure, if necessary, with the step length doubling each time, the minimum of $y(\lambda)$ is eventually 'boxed in' and the cubic interpolation starts.

It can be shown quite simply, using the calculus of variations, that of all functions $y(\lambda)$ which satisfy conditions (3.24), the cubic has the interesting property of being the 'smoothest', in the sense that it minimizes the integral

$$\int_0^\alpha \left(\frac{d^2y}{d\lambda^2}\right)^2 d\lambda. \tag{3.25}$$

For the Euler equation corresponding to (3.25) is

$$\frac{d^2}{d\lambda^2}\left(\frac{\partial F}{\partial y''}\right) = 0,$$

where $y'' = d^2y/d\lambda^2$ and $F = (y'')^2$. Thus

$$\frac{d^4y}{d\lambda^4} = 0,$$

giving a cubic for $y(\lambda)$. The fact that the cubic possesses this property was one of the reasons for Davidon's choice of cubic interpolation.

The interpolation formula for stage (c) is obtained as follows. Assume that the approximating cubic is given by

$$y(\lambda) \equiv f_0 + G_0\lambda + y_2\lambda^2 + y_3\lambda^3, \tag{3.26}$$

which already satisfies the first and third of conditions (3.24). Then

$$y'(\lambda) \equiv G_0 + 2y_2\lambda + 3y_3\lambda^2. \tag{3.27}$$

Setting $\lambda = \alpha$ in the identities (3.26) and (3.27) and using the second and fourth of conditions (3.24), we obtain two equations for y_2 and y_3:

$$\begin{aligned} \alpha^2 y_2 + \alpha^3 y_3 &= f_\alpha - f_0 - G_0\alpha, \\ 2\alpha y_2 + 3\alpha^2 y_3 &= G_\alpha - G_0. \end{aligned}$$

It is convenient to solve these equations for αy_2 and $\alpha^2 y_3$:

$$\alpha y_2 = -(G_0 + z), \tag{3.28}$$
$$\alpha^2 y_3 = \tfrac{1}{3}(G_0 + G_\alpha + 2z), \tag{3.29}$$

where

$$z = \frac{3}{\alpha}(f_0 - f_\alpha) + G_0 + G_\alpha. \tag{3.30}$$

The identity (3.27) now becomes

$$y'(\lambda) \equiv G_0 - 2(G_0 + z)\frac{\lambda}{\alpha} + (G_0 + G_\alpha + 2z)\frac{\lambda^2}{\alpha^2}. \tag{3.31}$$

To find λ_m, we set $y'(\lambda) = 0$ in (3.31) and solve for λ/α:

$$\begin{aligned} \frac{\lambda_m}{\alpha} &= \frac{G_0 + z \pm [(G_0 + z)^2 - G_0(G_0 + G_\alpha + 2z)]^{1/2}}{G_0 + G_\alpha + 2z} \\ &= \frac{G_0 + z \pm w}{G_0 + G_\alpha + 2z}, \end{aligned} \tag{3.32}$$

where

$$w = (z^2 - G_0 G_\alpha)^{1/2}. \tag{3.33}$$

The ambiguity in sign in equation (3.32) is resolved by considering the sign of $y''(\lambda_m)$. From (3.31) and (3.32), we obtain

$$\begin{aligned} y''(\lambda_m) &= -\frac{2}{\alpha}(G_0 + z) + 2(G_0 + G_\alpha + 2z)\frac{\lambda_m}{\alpha^2} \\ &= \pm 2w/\alpha \\ &> 0 \text{ for a minimum.} \end{aligned}$$

Hence we must take the + sign in equation (3.32), giving

$$\frac{\lambda_m}{\alpha} = \frac{G_0 + z + w}{G_0 + G_\alpha + 2z}. \tag{3.34}$$

Equation (3.34) solves the problem of finding the value of λ for which $y(\lambda)$ is a minimum. Davidon found, however, that for greater numerical accuracy it is preferable to use the equivalent formula

$$\frac{\lambda_m}{\alpha} = 1 - \frac{G_\alpha + w - z}{G_\alpha - G_0 + 2w}. \tag{3.35}$$

To show that equations (3.34) and (3.35) are equivalent, we have to show that

$$\frac{G_0 + z + w}{G_0 + G_\alpha + 2z} = \frac{-G_0 + w + z}{G_\alpha - G_0 + 2w};$$

but this result follows immediately by cross-multiplying and using equation (3.33).

Given a point $\mathbf{x}_k$ and a direction of search $\mathbf{d}$, where $\mathbf{d}$ need not be a unit vector, Davidon's cubic interpolation method for minimizing the differentiable function $f(\mathbf{x})$ on the line $\mathbf{x} = \mathbf{x}_k + \lambda\mathbf{d}$ is as follows:

1. Evaluate $f_0 = f(\mathbf{x}_k)$ and $G_0 = [\mathbf{g}(\mathbf{x}_k)]'\mathbf{d}$. Check that $G_0 < 0$. In equation (3.23), choose κ and f_e, and hence determine α. The value $\kappa = 2$ is normally used.
2. Evaluate $f_\alpha = f(\mathbf{x}_k + \alpha\mathbf{d})$ and $G_\alpha = [\mathbf{g}(\mathbf{x}_k + \alpha\mathbf{d})]'\mathbf{d}$.
3. If $G_\alpha > 0$, or if $f_\alpha > f_0$, proceed to rule 5. Otherwise go to rule 4.
4. Replace α by 2α, evaluate the new f_α and G_α and return to rule 3.
5. Interpolate in the interval $[0, \alpha]$ for λ_m, using equation (3.35). In this equation, w and z are given by equations (3.33) and (3.30), respectively.
6. Return to rule 5 to repeat the interpolation in the smaller interval $[0, \lambda_m]$ or $[\lambda_m, \alpha]$, according as

$$[\mathbf{g}(\mathbf{x}_k + \lambda_m\mathbf{d})]'\mathbf{d} \geqslant \text{ or } < 0.$$

Stop when the interval of interpolation has decreased to some prescribed value.

Example 3.8

Use Davidon's cubic interpolation method to solve the problem of Example 3.7, *page* 96.

Solution

$$f(\mathbf{x}) \equiv x_1^4 - x_1^3 x_2 - x_1^2 x_2 x_3 + x_1 x_2 x_3 x_4,$$
$$\mathbf{g}(\mathbf{x}) \equiv \nabla f(\mathbf{x}) \equiv [4x_1^3 - 3x_1^2 x_2 - 2x_1 x_2 x_3 + x_2 x_3 x_4,$$
$$-x_1^3 - x_1^2 x_3 + x_1 x_3 x_4, \ -x_1^2 x_2 + x_1 x_2 x_4, \ x_1 x_2 x_3].$$

Iteration 1. The current point is $\mathbf{x}_k = [0, -1, -2, -3]$ and $f_0 = f(\mathbf{x}_k) = 0$. Also,

$$\mathbf{g}(\mathbf{x}_k) = [-6, 0, 0, 0],$$
$$G_0 = [\mathbf{g}(\mathbf{x}_k)]'\mathbf{d} = [-6, 0, 0, 0]'[1, 2, 3, 4] = -6.$$

We choose $\kappa = 2$ and $f_e = -0{\cdot}5$. Then

$$\alpha = \min\left\{2, \frac{-2(0 + 0{\cdot}5)}{-6}\right\} = \frac{1}{6}$$

$$f_\alpha = f(\mathbf{x}_k + \alpha\mathbf{d}) = f(0{\cdot}1667, -0{\cdot}6667, -1{\cdot}5, -2{\cdot}3333) = -0{\cdot}4128,$$
$$G_\alpha = [\mathbf{g}(\mathbf{x}_k + \alpha\mathbf{d})]'\mathbf{d}$$
$$= [-2{\cdot}5926, 0{\cdot}6204, 0{\cdot}2778, 0{\cdot}1667]'[1, 2, 3, 4] = 0{\cdot}1481.$$

Since $G_\alpha > 0$ we proceed to rule 5. We find

$$z = 1{\cdot}5787, \quad w = 1{\cdot}8388, \quad \lambda_m = 0{\cdot}1597.$$

Iteration 2. The new current point is

$$\mathbf{x}_k + \lambda_m\mathbf{d} = [0{\cdot}1597, -0{\cdot}6805, -1{\cdot}5208, -2{\cdot}3610]$$

and $f(\mathbf{x}_k + \lambda_m\mathbf{d}) = -0{\cdot}4133$. Following rule 6, we evaluate

$$[\mathbf{g}(\mathbf{x}_k + \lambda_m\mathbf{d})]'\mathbf{d} = -0{\cdot}005747,$$

and since this value is negative we repeat the interpolation over the smaller interval $[\lambda_m, \alpha]$.

It is convenient to use the original notation in the smaller interval, and so we replace $\mathbf{x}_k + \lambda_m \mathbf{d}$ by $\mathbf{x}_k$ and re-define the interval as $[0, \alpha]$, where the length $(\alpha - \lambda_m)$ of the interval has been replaced by α. We find that $\alpha = 0{\cdot}006925$, correct to four significant figures. We then have

$$f_0 = -0{\cdot}4133, \qquad G_0 = -0{\cdot}005747,$$
$$f_\alpha = -0{\cdot}4128, \qquad G_\alpha = 0{\cdot}1481.$$

Proceeding as before, we find

$$z = -0{\cdot}07314, \ w = 0{\cdot}07874, \ \lambda_m = 0{\cdot}0002525.$$

Iteration 3. The new current point is

$$\mathbf{x}_k + \lambda_m \mathbf{d} = [0{\cdot}1600, -0{\cdot}6800, -1{\cdot}5200, -2{\cdot}3600]$$

and $f(\mathbf{x}_k + \lambda_m \mathbf{d}) = -0{\cdot}4133$. Since there is no change in function value between iterations 2 and 3, we conclude that to the present accuracy the results of iteration 3 are optimal. These results should be compared with those of Example 3.7.

SUMMARY

Since the advent of electronic computers in 1945, it has become feasible to solve optimization problems by means of tedious repetitive steps. The simplest type of algorithm to which this description may be applied is the direct search method, which uses function evaluations and comparisons together with a strategy for determining the next trial point. Sections 3.2 to 3.7 of the present chapter contain a representative selection of the many ingenious direct search methods that have been devised. The method of Hooke and Jeeves is a widely used general purpose method of this type. Historically, it was preceded by Rosenbrock's method[71] and followed by the method of Davies, Swann and Campey,[74] the latter being a development of Rosenbrock's method; the original references should be consulted for details of these methods. Several direct search methods use geometrical designs, of which the rectangular grid is the simplest though least efficient. The simplex methods of Spendley, Hext and Himsworth and of Nelder and Mead are much more efficient algorithms in this class. Fibonacci and Golden Section searches are simple and efficient, though they are limited to unimodal functions. The chapter ends with two well-known linear search techniques—the quadratic interpolation method of Powell and the cubic interpolation method of Davidon. The former uses function values only, while the latter requires the evaluation of the function and its derivatives. Davidon's method is to be preferred if the derivatives can be evaluated reasonably simply.

EXERCISES

1. Using not more than thirty-five function evaluations in each case, find

the maximum value of

$$f(x,y) \equiv (2x-1)\, y^2 e^{-(x+y)}$$

in the region $x \in [0,2]$, $y \in [0,3]$, by means of

(a) a grid search without feedback,

(b) a grid search with feedback,

(c) a random search.

2. Construct flow diagrams for

(a) the method of Hooke and Jeeves,

(b) the method of Spendley, Hext and Himsworth,

(c) the method of Nelder and Mead.

3. Use Hooke and Jeeves' method to minimize

$$f(\mathbf{x}) \equiv 5x_1^2 - 16x_1x_2 + 13x_2^2 + 10x_1 - 16x_2,$$

taking $\mathbf{b}_1 = [0,0]$ as the initial point, $h_1 = h_2 = 2$ as initial step lengths and $h_1 = h_2 < 1$ as the stopping condition. Repeat the working with $\mathbf{b}_1 = [4,2]$ and comment on the result.

4. Hooke and Jeeves' method may be used to solve constrained optimization problems by the simple device of regarding as a failure any exploratory or pattern move which takes the current point out of the feasible region. Use this technique to minimize

$$f(\mathbf{x}) \equiv x_1^2 - x_1x_2 + x_2^2 - x_1 + x_2,$$

subject to

$$x_1 + x_2 \geqslant 1.$$

Take $\mathbf{b}_1 = [2,2]$ as the initial point and $h_1 = h_2 = 0{\cdot}2$ as initial step lengths. Stop when $h_1 = h_2 < 0{\cdot}1$.

5. Use the method of Hooke and Jeeves to minimize the objective function of Exercise 3, subject to the constraint

$$x_1 + x_2 \geqslant 1.$$

Take $\mathbf{b}_1 = [\frac{1}{2},\frac{1}{2}]$ as the initial point, $h_1 = h_2 = 1$ as initial step lengths and $h_1 = h_2 < \frac{1}{8}$ as the stopping condition. Also, solve this problem exactly by the method of Section 1.4.

6. If the vertices of a regular simplex in E^2 are the points $[0,0]$, $[p,q]$ and $[q,p]$, and if its sides are of unit length, show that

$$p,q = (\sqrt{3} \pm 1)/2\sqrt{2}.$$

More generally, if the vertices of a regular simplex in E^n are the points $[0,0,....,0]$, $[p,q,q,....,q]$, $[q,p,q,....,q]$, $[q,q,p,q,....,q]$, ..., $[q,....,q,p]$, and if its sides are of unit length, show that

$$p = (n-1+\sqrt{n+1})/n\sqrt{2}, \quad q = (\sqrt{n+1}-1)/n\sqrt{2}.$$

7. Use the method of Spendley, Hext and Himsworth, either graphically or analytically, to minimize

$$f(\mathbf{x}) = 4x_1^2 - x_1 x_2 + x_2^2 - 15x_1,$$

starting with the simplex of Exercise 6 and stopping immediately before the third contraction. Estimate the maximum error in your answer.

8. Repeat Exercise 7 using the method of Nelder and Mead and determine the standard deviation of the function values at the vertices of the final simplex.

9. Use Fibonacci search to maximize $x\cos x$ over the interval $[0, \frac{\pi}{2}]$, with an error in x^* of not more than 0·001.

10. Using only ten function evaluations, find as accurately as possible the value of $x \in [-1{\cdot}9, 0{\cdot}9]$ which minimizes $(4x - 7)/(x^2 + x - 2)$.

11. Repeat Example 3.2, to find x^* with an error of *less than* 0·05, by following the method given in the proof of Theorem 3.1, with $\varepsilon = 0{\cdot}2$ in equation (3.7).

12. Find the minimum value of the function

$$\phi(X) \equiv X^4 - 2000X^3 + 100{,}000X$$

defined on the set of points $X = 1, 2, \ldots, 2000$. Deduce the minimum value of the function

$$f(x) \equiv x^4 - 20x^3 + 0{\cdot}1x$$

defined on the set of points $x = 0{\cdot}01, 0{\cdot}02, \ldots, 20{\cdot}00$.

13. Repeat Exercise 9 using Golden Section search.

14. Use (a) Powell's quadratic interpolation method, (b) Davidon's cubic interpolation method, to find the minimum value of

$$f(\mathbf{x}) \equiv x_1^4 - 3x_1 x_2 - 2x_2^2$$

in the negative gradient direction from the point $\mathbf{x}_1 = [1, 2]$, with an error in $|\mathbf{x}^*|$ not exceeding 0·01.

15. It is known from elementary geometry that the tangent to the parabola $y = kx^2$ at the point $[x_0, y_0]$ meets the x-axis at the point $[x_0/2, 0]$. Use this result to give a geometric interpretation of the expression $-2(f_0 - f_e)/G_0$ in equation (3.23).

16. Attempt to solve the problem of Example 3.7 starting from the point $\mathbf{x}_k = [1, 1, 1, 1]$ and using $h = 1{\cdot}0$, $H = 2{\cdot}0$, $\varepsilon = 0{\cdot}001$. Explain the result.

17. Use any method of this chapter to solve the simultaneous equations

$$\begin{aligned} x_1 - 2x_2 + 3x_3 &= 2, \\ 3x_1 - 2x_2 + x_3 &= 7, \\ x_1 + x_2 - x_3 &= 1. \end{aligned}$$

18. Construct a quadratic interpolation algorithm for the minimization of $f(\mathbf{x})$ along the line $\mathbf{x} = \mathbf{x}_k + \lambda\mathbf{d}$ where, in the notation of Section 3.9, the given values are f_0, G_0 and f_α.

CHAPTER 4

Gradient Methods for Unconstrained Optimization

4.1 INTRODUCTION

Gradient methods for finding a local maximum or minimum of an unconstrained function $f(\mathbf{x})$ are based ultimately on the simple fact that $f(\mathbf{x})$ increases or decreases in the direction $\mathbf{d}$ according as the directional derivative $[\nabla f(\mathbf{x})]'\mathbf{d}$ is positive or negative. The following questions immediately arise:

(a) Given a current point $\mathbf{x}_k$, can $\nabla f(\mathbf{x}_k)$ be evaluated without undue computational effort?

(b) How should the direction $\mathbf{d}$ be chosen?

(c) How large a step should be taken in the direction $\mathbf{d}$ from one current point to the next?

Some of the many answers to these questions provide the subject matter of this chapter. Without loss of generality, we shall consider only the minimizing problem.

We shall first consider gradient methods that require the evaluation of derivatives; these methods will be presented in order of increasing complexity. Typical of this class are the Davidon–Fletcher–Powell method (Section 4.4) and the Fletcher–Reeves method (Section 4.7). Both depend on the properties of conjugate directions (Section 4.6) and both have been adapted successfully to solve optimal control problems.[1,52]

Gradient methods that do not require the evaluation of derivatives also depend on the properties of conjugate directions. Such methods are used when it is not possible to find analytic expressions for the derivatives and also when these evaluations would be unduly difficult, inaccurate or time-consuming. Powell's method (Section 4.9) is the best-known method of this type.

Much of the theory of this chapter is valid only for the case where $f(\mathbf{x})$ is a quadratic function of $\mathbf{x}$. It must be emphasized, however, that gradient methods are also applicable to more general nonlinear functions. Quadratic functions are used for the theoretical development mainly because they lead to relatively simple theorems concerning the properties of the various methods. Corres-

TEST FUNCTIONS

ponding theorems for general nonlinear functions tend to be much more difficult to formulate and prove. It is not surprising, therefore, that many important properties of gradient methods have been discovered from practical experience and by the use of standard test functions.

Three well-known test functions which are in common use for the comparison of different optimization techniques are given below. All the functions are to be minimized, starting at the point $\mathbf{x}_1$.

Rosenbrock[71]

$$f(\mathbf{x}) \equiv 100(x_2 - x_1{}^2)^2 + (1 - x_1)^2;$$
$$\mathbf{x}_1 = [-1{\cdot}2, 1{\cdot}0] \quad \mathbf{x}^* = [1{\cdot}0, 1{\cdot}0].$$

Geometrically, the function represents a deep parabolic valley, and hence frequent changes of search direction are necessary.

Powell[61]

$$f(\mathbf{x}) \equiv (x_1 + 10x_2)^2 + 5(x_3 - x_4)^2 + (x_2 - 2x_3)^4 + 10(x_1 - x_4)^4;$$
$$\mathbf{x}_1 = [3, -1, 0, 1], \quad \mathbf{x}^* = [0, 0, 0, 0].$$

This is a quartic function with a Hessian matrix of rank 2. The function cannot therefore be approximated by a quadratic in the neighbourhood of the minimizing point.

Fletcher and Powell[36]

$$f(\mathbf{x}) \equiv 100[x_3 - 10\theta(x_1, x_2)]^2 + [\sqrt{x_1{}^2 + x_2{}^2} - 1]^2 + x_3{}^2,$$

where

$$\left.\begin{aligned}\cos 2\pi\theta(x_1, x_2) &= x_1/\sqrt{x_1{}^2 + x_2{}^2},\\ \sin 2\pi\theta(x_1, x_2) &= x_2/\sqrt{x_1{}^2 + x_2{}^2},\end{aligned}\right\} -\frac{\pi}{2} < 2\pi\theta \leqslant \frac{3\pi}{2};$$
$$\mathbf{x}_1 = [-1, 0, 0], \quad \mathbf{x}^* = [1, 0, 0].$$

Geometrically, the function represents a helical valley—a severe test for any optimization technique.

4.2 METHOD OF STEEPEST DESCENT

This method is not recommended for general use, but is included because of its simplicity and also because it helps to give some insight into the more sophisticated methods. In common with all gradient methods, the method of steepest descent (ascent for maximizing problems) is iterative, proceeding from an initial approximation $\mathbf{x}_1$ for the minimizing point to successive points $\mathbf{x}_2, \mathbf{x}_3, \ldots$, until some stopping condition is satisfied.

Given the current point $\mathbf{x}_k$, the point $\mathbf{x}_{k+1}$ is obtained by a linear search in the direction $\mathbf{d}_k$, where

$$\mathbf{d}_k = -\nabla f(\mathbf{x}_k) = -\mathbf{g}(\mathbf{x}_k) = -\mathbf{g}_k,$$

i.e. $\mathbf{d}_k$ is the negative gradient vector at the point $\mathbf{x}_k$. It is well known that this direction is the direction from $\mathbf{x}_k$ in which the initial rate of decrease of $f(\mathbf{x})$

is greatest. Thus the sequence $\{\mathbf{x}_k\}$ is defined by

$$\mathbf{x}_{k+1} = \mathbf{x}_k + \lambda_k^* \mathbf{d}_k = \mathbf{x}_k - \lambda_k^* \mathbf{g}_k \quad (k=1,2,....), \tag{4.1}$$

where $\mathbf{x}_1$ is given and λ_k^* is determined by the linear search, so that $\mathbf{x}_{k+1}$ minimizes $f(\mathbf{x})$ in the direction $-\mathbf{g}_k$ from $\mathbf{x}_k$. The search in E^n for a minimum of $f(x_1,....,x_n)$ is therefore reduced to a sequence of linear searches.

Alternatively, and more simply, λ_k can remain fixed for all k. In this case, the linear searches are not required, though the choice of step length is a compromise between accuracy and efficiency.

It is essential that the sequence $\{\mathbf{x}_k\}$ of equation (4.1) should converge to the desired optimal point. The following theorem provides sufficient conditions for convergence in the case where the minimum cannot be found in a finite number of iterations.

***Theorem* 4.1**

If the limit of the sequence $\{\mathbf{x}_k\}$ *of* (4.1) *is* $\mathbf{x}^*$ *and if* $f(\mathbf{x}) \in C_1$ *for all* $\mathbf{x}$ *in a suitable neighbourhood of* $\mathbf{x}^*$, *then* $f(\mathbf{x})$ *has a local minimum at* $\mathbf{x} = \mathbf{x}^*$.

Proof

The proof is by contradiction. Assume that $f(\mathbf{x}^*)$ is not a local minimum of $f(\mathbf{x})$. Then $[\mathbf{g}(\mathbf{x}^*)]^2 > 0$. Since $\mathbf{g}(\mathbf{x})$ is continuous, there exist numbers $\mu > 0$ and $\lambda > 0$ such that

$$\text{(i)} \quad [\mathbf{g}(\mathbf{x})]'\mathbf{g}(\mathbf{y}) \geqslant \mu > 0 \tag{4.2}$$

for all $\mathbf{x}$ and $\mathbf{y}$ in some neighbourhood of $\mathbf{x}^*$;

(ii) if $\mathbf{x}_k$ is a point in this neighbourhood then so is $\mathbf{x}_{k+1}$, where

$$\mathbf{x}_{k+1} = \mathbf{x}_k - \lambda \mathbf{g}(\mathbf{x}_k). \tag{4.3}$$

Consider the change in function value as we move from $\mathbf{x}_k$ to $\mathbf{x}_{k+1}$. By the first mean value theorem,

$$f(\mathbf{x}_{k+1}) = f(\mathbf{x}_k) - \lambda[\mathbf{g}(\mathbf{x}_k)]' \{\mathbf{g}[\mathbf{x}_k - \theta\lambda\mathbf{g}(\mathbf{x}_k)]\}, \quad 0 \leqslant \theta \leqslant 1. \tag{4.4}$$

In equation (4.4), the coefficient of $-\lambda$ is at least μ, from (4.2) and (4.3). Thus, in moving from $\mathbf{x}_k$ to $\mathbf{x}_{k+1}$, the value of $f(\mathbf{x})$ is decreased by at least $\lambda\mu$. But there is an infinite number of terms of the sequence $\{\mathbf{x}_k\}$ in any neighbourhood of $\mathbf{x}^*$, since

$$\lim_{k \to \infty} \mathbf{x}_k = \mathbf{x}^*.$$

Hence, by repeating the above argument for successive values of k, we find that $f(\mathbf{x}^*) \to -\infty$, which contradicts the fact that $\mathbf{g}(\mathbf{x}^*)$ exists. The original assumption that $f(\mathbf{x}^*)$ is not a local minimum of $f(\mathbf{x})$ is therefore false, and the theorem is proved.

In practice, the *rate* of convergence of the sequence $\{\mathbf{x}_k\}$ is important. A major disadvantage of the method of steepest descent is that no account is taken of the second derivatives of $f(\mathbf{x})$ and yet the curvature of the function—which determines its behaviour near the minimum—depends on these derivatives. This means that the rate of convergence of the sequence $\{\mathbf{x}_k\}$ can be very slow. The Newton–Raphson method of the next section overcomes this disadvantage.

4.3 THE NEWTON–RAPHSON METHOD

For analytic purposes, the simplest function (of any number of variables) with a strong minimum is a quadratic with a positive definite Hessian matrix. The basic idea of the Newton–Raphson method is to approximate the given function $f(\mathbf{x})$ in each iteration by such a quadratic function and then to move the current point to the turning point of the quadratic.

Assuming that $f(\mathbf{x}) \in C_2$ in a suitable neighbourhood of the current point $\mathbf{x}_k$, we evaluate $f(\mathbf{x})$ and its first and second derivatives at $\mathbf{x} = \mathbf{x}_k$, and find a quadratic function $y_k(\mathbf{x})$ which matches these values at $\mathbf{x} = \mathbf{x}_k$. This gives

$$y_k(\mathbf{x}) \equiv \tfrac{1}{2}(\mathbf{x} - \mathbf{x}_k)' \mathbf{G}_k(\mathbf{x} - \mathbf{x}_k) + (\mathbf{x} - \mathbf{x}_k)' \mathbf{g}_k + f(\mathbf{x}_k),$$

where $\mathbf{G}_k$ is the Hessian matrix and $\mathbf{g}_k$ the gradient vector of $f(\mathbf{x})$, both evaluated at $\mathbf{x} = \mathbf{x}_k$.

Suppose that $y_k(\mathbf{x})$ takes its minimum value at $\mathbf{x} = \mathbf{x}_m$. Then $\nabla y_k(\mathbf{x}_m) = 0$, i.e.

$$\mathbf{G}_k(\mathbf{x}_m - \mathbf{x}_k) + \mathbf{g}_k = 0,$$

which yields

$$\mathbf{x}_m = \mathbf{x}_k - \mathbf{G}_k^{-1} \mathbf{g}_k.$$

The Newton–Raphson method uses $\mathbf{x}_m$ as the next current point, giving the iterative formula

$$\mathbf{x}_{k+1} = \mathbf{x}_k - \mathbf{G}_k^{-1} \mathbf{g}_k \qquad (k = 1, 2, \ldots). \tag{4.5}$$

A variant of equation (4.5) that is often used, and which is usually superior, is

$$\mathbf{x}_{k+1} = \mathbf{x}_k - \lambda_k^* \mathbf{G}_k^{-1} \mathbf{g}_k \quad (k = 1, 2, \ldots), \tag{4.6}$$

where λ_k^* is determined by a linear search from $\mathbf{x}_k$ in the direction $-\mathbf{G}_k^{-1}\mathbf{g}_k$.

Greater numerical accuracy is attained if the variables x_j are scaled so that the Hessian matrices $\mathbf{G}_k$ are approximately equal to the unit matrix—this is true of all algorithms which make use of Hessian matrices or their inverses. Normally, the variables are scaled only once, at the beginning of the calculation.

The convergence of both forms, (4.5) and (4.6), of the Newton–Raphson method is rapid when $\mathbf{x}_k$ is near the optimal point $\mathbf{x}^*$. Whichever form is used, both $|\mathbf{x}_{k+1} - \mathbf{x}_k|$ and $|\mathbf{g}_k|$ should be tested in a convergence criterion. The success of the method depends on the direction $-\mathbf{G}_k^{-1}\mathbf{g}_k$ being a direction of descent. Thus we require

$$(-\mathbf{G}_k^{-1}\mathbf{g}_k)' \mathbf{g}_k < 0,$$

i.e.

$$\mathbf{g}_k' \mathbf{G}_k^{-1} \mathbf{g}_k > 0, \tag{4.7}$$

which is satisfied at all points for which $\mathbf{g}_k \neq \mathbf{0}$ if $\mathbf{G}_k$ is positive definite. Unfortunately, if $\mathbf{x}_k$ is not close to $\mathbf{x}^*$, it may happen that $\mathbf{G}_k$ is not positive definite; the method may fail to converge in this case.

The following example shows that when $\mathbf{G}_k$ is positive semidefinite the direction of search may become orthogonal to the local gradient direction—with disastrous results.

Example 4.1

Minimize

$$f(\mathbf{x}) \equiv x_1{}^4 - 3x_1x_2 + (x_2+2)^2,$$

starting at the point $\mathbf{x}_1 = [0,0]$.

Solution

The gradient vector and Hessian matrix of $f(\mathbf{x})$ are

$$\mathbf{g}(\mathbf{x}) \equiv [4x_1{}^3 - 3x_2, -3x_1 + 2(x_2+2)], \quad \mathbf{G}(\mathbf{x}) \equiv \begin{pmatrix} 12x_1{}^2 & -3 \\ -3 & 2 \end{pmatrix}$$

Therefore

$$\mathbf{g}_1 = [0,4], \quad \mathbf{G}_1 = \begin{pmatrix} 0 & -3 \\ -3 & 2 \end{pmatrix},$$

$$\mathbf{G}_1{}^{-1} = -\frac{1}{9}\begin{pmatrix} 2 & 3 \\ 3 & 0 \end{pmatrix}, \quad -\mathbf{G}_1{}^{-1}\mathbf{g}_1 = [4/3, 0].$$

Searching in the direction $[4/3,0]$ from the point $[0,0]$, i.e. following the method of equation (4.6), we find

$$f(\mathbf{x}_2) = f\left(\frac{4}{3}\lambda_1, 0\right) = \frac{256}{81}\lambda_1{}^4 + 4,$$

which is a minimum when $\lambda_1 = 0$. Thus no progress can be made from the initial point $\mathbf{x}_1$, although $\mathbf{g}_1 \neq \mathbf{0}$. Also, using equation (4.5), we find

$$f(\mathbf{x}_2) = \frac{256}{81} + 4 > f(\mathbf{x}_1),$$

and so both versions of the method fail. An obvious remedy in both cases is to move some distance in the negative gradient direction from the current point $\mathbf{x}_1$ before resuming the Newton–Raphson search.

Comparison of the iterative formulae (4.1) and (4.6) shows that the method of steepest descent and the Newton–Raphson method are identical when $\mathbf{G}_k{}^{-1} = \mathbf{I}$, the unit matrix. This observation leads to a whole new class of gradient methods, known as *quasi-Newton* or *variable metric methods*. In these methods, the matrix $\mathbf{G}_k{}^{-1}$ of equation (4.6), which may be difficult to evaluate, is replaced by a positive definite symmetric matrix $\mathbf{H}_k$ which is updated in each iteration without the need for matrix inversion. The resulting iterative formula is

$$\mathbf{x}_{k+1} = \mathbf{x}_k - \lambda_k{}^* \mathbf{H}_k \mathbf{g}_k \quad (k = 1, 2, \ldots.), \tag{4.8}$$

with $\mathbf{x}_1$ and $\mathbf{H}_1$ given. There are, of course, many choices available for $\mathbf{H}_k$.

The theory and applications of quasi-Newton methods have been considered by many authors, e.g. Broyden,[11–15] Powell,[64–66] Pearson,[60] Gill and Murray[39] and Fletcher.[34] In the next section, we describe in detail one of the best-known methods of this type, namely, the Davidon–Fletcher–Powell method.

4.4 THE DAVIDON–FLETCHER–POWELL METHOD

This method, often called the DFP method, was originally devised by Davidon[24] in 1959 and was later improved by Fletcher and Powell[36] in 1963. It uses the iterative formula (4.8) with $\mathbf{H}_1 = \mathbf{I}$; thus the first step (for a minimizing problem) is in the negative gradient direction. The slow convergence of the method of steepest descent near the optimal point $\mathbf{x}^*$ is overcome by choosing the sequence $\{\mathbf{H}_k\}$ in such a way that $\mathbf{H}_k$ becomes approximately equal to $\mathbf{G}_k{}^{-1}$ as $\mathbf{x}_k$ approaches $\mathbf{x}^*$. In fact, if $f(\mathbf{x})$ is a *quadratic* function of n variables x_j, with Hessian matrix $\mathbf{G}$, then $\mathbf{H}_{n+1} = \mathbf{G}^{-1}$; also, the exact minimum is reached in, at most, n iterations. Near the initial point the DFP method therefore resembles the method of steepest descent in its initial rapid reduction of the value of $f(\mathbf{x})$; near the optimal point it resembles the Newton–Raphson method in its ultimate fast convergence.

We shall now discuss the theory of the DFP method for the case where $f(\mathbf{x})$ is a quadratic function. The fact that most of the theoretical results are valid only for quadratic objective functions does not, of course, prevent the method being applied iteratively to non-quadratic objective functions. A variant of the DFP method which makes explicit use of non-quadratic properties of the objective function has been devised by Biggs.[7]

Consider the problem of minimizing the quadratic function

$$f(\mathbf{x}) \equiv \tfrac{1}{2}\mathbf{x}'\mathbf{G}\mathbf{x} + \mathbf{b}'\mathbf{x} + c, \tag{4.9}$$

where $\mathbf{G}$ is positive definite and is assumed, without loss of generality, to be symmetric. The gradient vector at the point $\mathbf{x} = \mathbf{x}_k$ is

$$\mathbf{g}(\mathbf{x}_k) = \mathbf{g}_k = \nabla f(\mathbf{x}_k) = \mathbf{G}\mathbf{x}_k + \mathbf{b}. \tag{4.10}$$

Hence if $\mathbf{x}^*$ is the optimal point, given by $\mathbf{g}(\mathbf{x}^*) = 0$, then

$$\mathbf{x}^* = \mathbf{x}_k - \mathbf{G}^{-1}\mathbf{g}_k. \tag{4.11}$$

However, as stated above, $\mathbf{G}^{-1}$ is replaced in each iteration by a positive definite symmetric matrix $\mathbf{H}_k$ with $\mathbf{H}_1 = \mathbf{I}$, and this leads to the iterative formula (4.8).

It is convenient at this stage to write down the steps in the kth iteration of the DFP method for the minimization of $f(\mathbf{x})$. The initial point is $\mathbf{x}_1$, and the current point at the start of the kth iteration is $\mathbf{x}_k$.

1. Set

$$\mathbf{d}_k = -\mathbf{H}_k \mathbf{g}_k, \tag{4.12}$$

with $\mathbf{H}_1 = \mathbf{I}$. Then $\mathbf{d}_k$ is the direction of search from the current point $\mathbf{x}_k$.

2. Perform a linear search to find λ_k^* (>0), where λ_k^* is the value of λ_k that minimizes $f(\mathbf{x}_k + \lambda_k \mathbf{d}_k)$.

3. Set

$$\boldsymbol{\sigma}_k = \lambda_k^* \mathbf{d}_k. \tag{4.13}$$

4. Set

$$\mathbf{x}_{k+1} = \mathbf{x}_k + \boldsymbol{\sigma}_k, \tag{4.14}$$

giving the new current point.

5. Evaluate $f(\mathbf{x}_{k+1})$ and $\mathbf{g}_{k+1}$, noting that $\mathbf{g}_{k+1}$ is orthogonal to $\boldsymbol{\sigma}_k$, i.e.

$$\boldsymbol{\sigma}_k' \mathbf{g}_{k+1} = 0, \tag{4.15}$$

for, at the point $\mathbf{x} = \mathbf{x}_{k+1}$, the vector $\boldsymbol{\sigma}_k$ is tangential to the level hypersurface $f(\mathbf{x}) = f(\mathbf{x}_{k+1})$ and $\mathbf{g}_{k+1}$ is normal to this hypersurface.

6. Set

$$\boldsymbol{\gamma}_k = \mathbf{g}_{k+1} - \mathbf{g}_k. \tag{4.16}$$

7. Set

$$\mathbf{H}_{k+1} = \mathbf{H}_k + \mathbf{A}_k + \mathbf{B}_k, \tag{4.17}$$

where

$$\mathbf{A}_k = \boldsymbol{\sigma}_k \boldsymbol{\sigma}_k' / \boldsymbol{\sigma}_k' \boldsymbol{\gamma}_k, \tag{4.18}$$

$$\mathbf{B}_k = \mathbf{H}_k \boldsymbol{\gamma}_k \boldsymbol{\gamma}_k' \mathbf{H}_k / \boldsymbol{\gamma}_k' \mathbf{H}_k \boldsymbol{\gamma}_k. \tag{4.19}$$

8. Set $k = k + 1$ and return to step 1.

9. Stop when either $|\mathbf{d}_k|$ or, as a safer alternative, every component of $\mathbf{d}_k$ is smaller than some prescribed amount. In practice, whichever of these tests is applied to $\mathbf{d}_k$ is also applied to $\boldsymbol{\sigma}_k$ and the calculations are terminated when either $\mathbf{d}_k$ or $\boldsymbol{\sigma}_k$ satisfies the chosen criterion. As a further safeguard, the value of $|\mathbf{g}_k|$ may be checked. Fletcher and Powell recommend that the calculations be continued for at least n iterations in order to avoid false minima.

Any one-dimensional search technique may be used in step 2, though it is usual to use Davidon's cubic interpolation method (Section 3.9) with $\kappa = 2$ in equation (3.23).

We shall develop the theory of the DFP method under three headings: (a) stability, (b) convergence and (c) the formula (4.17) for updating $\mathbf{H}_k$.

(a) Stability

An iterative method for the minimization of $f(\mathbf{x})$ is said to be *stable* if the value of $f(\mathbf{x})$ is reduced in each iteration. Following the argument which led to condition (4.7), the DFP method is seen to be stable if $\mathbf{H}_k$ is positive definite for all k.

***Theorem* 4.2**

In the DFP method, $\mathbf{H}_k$ *is positive definite for all* k.

Proof

The proof is inductive. First, $\mathbf{H}_1 = \mathbf{I}$, which is positive definite. Now assume that the theorem is true for $k = K$; we shall prove that it is true for $k = K + 1$.

In step 2, the direction of search is 'downhill', i.e. $\mathbf{g}_K'\mathbf{d}_K < 0$, and hence $\lambda_K^* > 0$. Define the vectors

$$\mathbf{p} = \mathbf{H}_K^{1/2}\xi \quad \text{and} \quad \mathbf{q} = \mathbf{H}_K^{1/2}\gamma_K,$$

where ξ is an arbitrary non-zero vector. The proof that $\mathbf{H}_K^{1/2}$ exists is left to the exercises. From equations (4.17) to (4.19), we find

$$\begin{aligned}
\xi'\mathbf{H}_{K+1}\xi &= \xi'\mathbf{H}_K\xi + \frac{(\xi'\sigma_K)^2}{\sigma_K'\gamma_K} - \frac{(\xi'\mathbf{H}_K\gamma_K)^2}{\gamma_K'\mathbf{H}_K\gamma_K} \\
&= \mathbf{p}^2 + \frac{(\xi'\sigma_K)^2}{\sigma_K'\gamma_K} - \frac{(\mathbf{p}'\mathbf{q})^2}{\mathbf{q}^2} \\
&= \frac{\mathbf{p}^2\mathbf{q}^2 - (\mathbf{p}'\mathbf{q})^2}{\mathbf{q}^2} + \frac{(\xi'\sigma_K)^2}{\sigma_K'\gamma_K} \\
&\geqslant \frac{(\xi'\sigma_K)^2}{\sigma_K'\gamma_K},
\end{aligned}$$

where we have used Schwartz's inequality. Now

$$\begin{aligned}
\sigma_K'\gamma_K &= \sigma_K'\mathbf{g}_{K+1} - \sigma_K'\mathbf{g}_K \\
&= -\sigma_K'\mathbf{g}_K, && \text{using equation (4.15)}, \\
&= \lambda_K^*\mathbf{g}_K'\mathbf{H}_K\mathbf{g}_K, && \text{using equations (4.12) and (4.13)}, \\
&> 0,
\end{aligned}$$

since $\lambda_K^* > 0$ and $\mathbf{H}_K$ is positive definite. Hence $\xi'\mathbf{H}_{K+1}\xi > 0$ for all non-zero ξ, i.e. $\mathbf{H}_{K+1}$ is positive definite, and the induction is complete.

As stated above, it follows immediately from Theorem 4.2 that the DFP method is stable. It should be noted that the result we have just obtained is entirely theoretical. It is possible in practice for $\mathbf{H}_k$ to become singular owing to numerical errors; the consequences of this are considered in Exercise 8 at the end of the chapter.

(*b*) *Convergence*

An iterative method for function minimization is said to have the property of *quadratic termination* if it reaches the exact optimal point of a quadratic function $f(\mathbf{x})$ in a finite number of iterations. We shall prove that the DFP method possesses this property and, moreover, that for a function of n variables it reaches the optimal point in, at most, n iterations. It should be remembered, however, that these statements are true only if the linear searches find the exact minima and if the accompanying arithmetic is also exact.

Assume that $f(\mathbf{x})$ is the quadratic function (4.9). The main part of the proof is contained in the following theorem.

***Theorem* 4.3**

If the DFP method is used to minimize the quadratic function (4.9), *with* $\mathbf{G}$ *positive definite and symmetric, then* $\mathbf{H}_{n+1} = \mathbf{G}^{-1}$.

Proof

The notation is given in equations (4.9) to (4.19). We need two preliminary results. First,

$$\gamma_k = \mathbf{g}_{k+1} - \mathbf{g}_k = \mathbf{G}(\mathbf{x}_{k+1} - \mathbf{x}_k) = \mathbf{G}\boldsymbol{\sigma}_k. \tag{4.20}$$

Secondly, using equations (4.20) and (4.17) to (4.19), we obtain

$$\begin{aligned} \mathbf{H}_{k+1}\mathbf{G}\boldsymbol{\sigma}_k &= \mathbf{H}_{k+1}\gamma_k \\ &= \mathbf{H}_k\gamma_k + \frac{\boldsymbol{\sigma}_k\boldsymbol{\sigma}_k'\gamma_k}{\boldsymbol{\sigma}_k'\gamma_k} - \frac{\mathbf{H}_k\gamma_k\gamma_k'\mathbf{H}_k\gamma_k}{\gamma_k'\mathbf{H}_k\gamma_k} \\ &= \boldsymbol{\sigma}_k. \end{aligned} \tag{4.21}$$

Equation (4.21) shows that $\boldsymbol{\sigma}_k$ is an eigenvector of $\mathbf{H}_{k+1}\mathbf{G}$ with an eigenvalue of unity. We shall prove the more general result that for $m = 2,\ldots,n+1$ the vectors $\boldsymbol{\sigma}_1,\ldots,\boldsymbol{\sigma}_{m-1}$ of equation (4.13) are linearly independent eigenvectors of $\mathbf{H}_m\mathbf{G}$ with eigenvalues of unity. The proof is by mathematical induction on m.

Consider the equations

$$\boldsymbol{\sigma}_k'\mathbf{G}\boldsymbol{\sigma}_l = 0 \qquad (1 \leqslant k < l < m) \tag{4.22}$$

and

$$\mathbf{H}_m\mathbf{G}\boldsymbol{\sigma}_k = \boldsymbol{\sigma}_k \qquad (1 \leqslant k < m). \tag{4.23}$$

Note that $m \geqslant 3$ in equations (4.22) and that $m \geqslant 2$ in equations (4.23). When $m = 2$, the single equation (4.23) becomes

$$\mathbf{H}_2\mathbf{G}\boldsymbol{\sigma}_1 = \boldsymbol{\sigma}_1, \tag{4.24}$$

which is true because of equation (4.21) with $k = 1$. When $m = 3$, the left-hand side of the single equation (4.22) becomes

$$\boldsymbol{\sigma}_1'\mathbf{G}\boldsymbol{\sigma}_2 = \boldsymbol{\sigma}_1'\mathbf{G}(-\lambda_2{}^*\mathbf{H}_2\mathbf{g}_2)$$

$$
\begin{aligned}
&= -\lambda_2{}^*\mathbf{g}_2{}'\mathbf{H}_2\mathbf{G}\boldsymbol{\sigma}_1, && \text{taking the transpose,}\\
&= -\lambda_2{}^*\mathbf{g}_2{}'\boldsymbol{\sigma}_1, && \text{using equation (4.24),}\\
&= 0, && \qquad (4.25)
\end{aligned}
$$

using equation (4.15) with $k = 1$.

Again when $m = 3$, equations (4.23) give

$$\mathbf{H}_3\mathbf{G}\boldsymbol{\sigma}_2 = \boldsymbol{\sigma}_2 \qquad \text{and} \qquad \mathbf{H}_3\mathbf{G}\boldsymbol{\sigma}_1 = \boldsymbol{\sigma}_1.$$

The first of these equations is true because of equation (4.21) with $k = 2$, and the left-hand side of the second equation is

$$
\begin{aligned}
\mathbf{H}_3\mathbf{G}\boldsymbol{\sigma}_1 &= \mathbf{H}_2\mathbf{G}\boldsymbol{\sigma}_1 + \frac{\boldsymbol{\sigma}_2\boldsymbol{\sigma}_2{}'\mathbf{G}\boldsymbol{\sigma}_1}{\boldsymbol{\sigma}_2{}'\boldsymbol{\gamma}_2} - \frac{\mathbf{H}_2\boldsymbol{\gamma}_2\boldsymbol{\gamma}_2{}'\mathbf{H}_2\mathbf{G}\boldsymbol{\sigma}_1}{\boldsymbol{\gamma}_2{}'\mathbf{H}_2\boldsymbol{\gamma}_2}\\
&= \boldsymbol{\sigma}_1, \qquad (4.26)
\end{aligned}
$$

using equations (4.20), (4.24) and (4.25)—the second and third terms on the right-hand side vanish separately. Thus equations (4.22) and (4.23) are true for $m = 3$. Now assume that they are true for $m = M$; we shall prove that they are true for $m = M + 1$.

From equation (4.10),

$$
\begin{aligned}
\mathbf{g}_M &= \mathbf{b} + \mathbf{G}\mathbf{x}_M\\
&= \mathbf{b} + \mathbf{G}(\mathbf{x}_{k+1} + \boldsymbol{\sigma}_{k+1} + \boldsymbol{\sigma}_{k+2} + \cdots + \boldsymbol{\sigma}_{M-1})\\
&= \mathbf{g}_{k+1} + \mathbf{G}(\boldsymbol{\sigma}_{k+1} + \cdots + \boldsymbol{\sigma}_{M-1}). \qquad (4.27)
\end{aligned}
$$

Using equations (4.15) and (4.22), this gives

$$\boldsymbol{\sigma}_k{}'\mathbf{g}_M = 0 \quad (1 \leqslant k < M). \qquad (4.28)$$

[In deriving equations (4.28), note that equation (4.15) is used in two different ways according as $k < M - 1$ or $k = M - 1$. When $k < M - 1$,

$$\boldsymbol{\sigma}_k{}'\mathbf{g}_{k+1} = 0 \quad \text{and} \quad \boldsymbol{\sigma}_k{}'\mathbf{G}\boldsymbol{\sigma}_{k+r} = 0 \quad (r = 1, \ldots, M-k-1), \qquad (4.29)$$

i.e. the scalar product of the right-hand side of equation (4.27) with $\boldsymbol{\sigma}_k$ vanishes. On the other hand, when $k = M - 1$, equation (4.15) gives directly

$$\boldsymbol{\sigma}_k{}'\mathbf{g}_M = \boldsymbol{\sigma}_{M-1}{}'\mathbf{g}_M = 0. \qquad (4.30)$$

Equations (4.28) follow by combining equations (4.29) and (4.30).]

From equations (4.23) and (4.28), we have

$$\boldsymbol{\sigma}_k{}'\mathbf{G}\mathbf{H}_M\mathbf{g}_M = \boldsymbol{\sigma}_k{}'\mathbf{g}_M = 0 \quad (1 \leqslant k < M),$$

so that

$$\boldsymbol{\sigma}_k{}'\mathbf{G}\boldsymbol{\sigma}_M = 0 \qquad (1 \leqslant k < M), \qquad (4.31)$$

because of equations (4.12) and (4.13). Now equations (4.31) can be written

$$\boldsymbol{\sigma}_k{}'\mathbf{G}\boldsymbol{\sigma}_l = 0 \qquad (1 \leqslant k < l < M + 1), \qquad (4.32)$$

where M is now included as a value of l. Note that equations (4.32) are simply equations (4.22) with $m = M + 1$.

From equations (4.20), (4.23) and (4.31), we obtain

$$\left.\begin{aligned} \gamma_M{}'\mathbf{H}_M\mathbf{G}\boldsymbol{\sigma}_k &= \gamma_M{}'\boldsymbol{\sigma}_k \\ &= \boldsymbol{\sigma}_M{}'\mathbf{G}\boldsymbol{\sigma}_k \\ &= 0 \qquad\qquad (1 \leqslant k < M). \end{aligned}\right\} \tag{4.33}$$

Hence

$$\begin{aligned} \mathbf{H}_{M+1}\mathbf{G}\boldsymbol{\sigma}_k &= \mathbf{H}_M\mathbf{G}\boldsymbol{\sigma}_k + \frac{\boldsymbol{\sigma}_M\boldsymbol{\sigma}_M{}'\mathbf{G}\boldsymbol{\sigma}_k}{\boldsymbol{\sigma}_M{}'\gamma_M} - \frac{\mathbf{H}_M\gamma_M\gamma_M{}'\mathbf{H}_M\mathbf{G}\boldsymbol{\sigma}_k}{\gamma_M{}'\mathbf{H}_M\gamma_M} \\ &= \mathbf{H}_M\mathbf{G}\boldsymbol{\sigma}_k, \end{aligned}$$

the second and third terms vanishing because of equations (4.33). It now follows from equations (4.23) that

$$\mathbf{H}_{M+1}\mathbf{G}\boldsymbol{\sigma}_k = \boldsymbol{\sigma}_k \qquad (1 \leqslant k < M). \tag{4.34}$$

Also, from equation (4.21) with $k = M$,

$$\mathbf{H}_{M+1}\mathbf{G}\boldsymbol{\sigma}_M = \boldsymbol{\sigma}_M. \tag{4.35}$$

Combining equations (4.34) and (4.35), we obtain

$$\mathbf{H}_{M+1}\mathbf{G}\boldsymbol{\sigma}_k = \boldsymbol{\sigma}_k \qquad (1 \leqslant k < M+1),$$

which are simply equations (4.23) with $m = M+1$.

We have now shown that if equations (4.22) and (4.23) are true for $m = M$ then they are true for $m = M+1$. It follows that equations (4.22) are true for $m = 3,\ldots,n+1$ and that equations (4.23) are true for $m = 2,\ldots,n+1$.

It will appear later (Section 4.6) that equations (4.22) imply that the directions $\boldsymbol{\sigma}_k, \boldsymbol{\sigma}_l$ $(1 \leqslant k < l < m)$ are, by definition, *mutually conjugate*, and thence by a very simple proof (Theorem 4.4) that the vectors $\boldsymbol{\sigma}_1,\ldots,\boldsymbol{\sigma}_{m-1}$ are linearly independent. Anticipating this result, equations (4.23) with $m = n+1$ imply that $\boldsymbol{\sigma}_1,\ldots,\boldsymbol{\sigma}_n$ are n linearly independent eigenvectors of $\mathbf{H}_{n+1}\mathbf{G}$ with eigenvalues of unity. It follows that $\mathbf{H}_{n+1}\mathbf{G}$ can be transformed into the unit matrix by a similarity transformation and hence that $\mathbf{H}_{n+1}\mathbf{G}$ *is* the unit matrix. Thus $\mathbf{H}_{n+1} = \mathbf{G}^{-1}$ and the theorem is proved.

Equations (4.28) with $M = n+1$ show that the DFP method finds the optimal point $\mathbf{x}^*$ of the quadratic function $f(\mathbf{x})$ in, at most, n iterations, for the n-component vector $\mathbf{g}_{n+1}$ must be orthogonal to each of the n linearly independent vectors $\boldsymbol{\sigma}_1,\ldots,\boldsymbol{\sigma}_n$, and this is only possible if $\mathbf{g}_{n+1} = \mathbf{0}$, i.e. if $\mathbf{x}_{n+1} = \mathbf{x}^*$.

The above discussion of convergence is based on the assumption of a quadratic objective function. Powell[64] has proved several theorems on the convergence of the DFP method for the case where $f(\mathbf{x})$ is not restricted to being a quadratic function.

(c) *Formula for updating* $\mathbf{H}_k$

So far, we have used equations (4.17) to (4.19) without explanation. It is now possible to interpret the formula (4.17) for updating $\mathbf{H}_k$. First, we shall prove that $\mathbf{A}_k$ is the term that makes $\mathbf{H}_k$ tend to $\mathbf{G}^{-1}$, in the sense that, for the quadratic objective function (4.9),

$$\mathbf{G}^{-1} = \sum_{k=1}^{n} \mathbf{A}_k. \tag{4.36}$$

To prove equation (4.36), let $\mathbf{S}$ be the matrix of column vectors $\boldsymbol{\sigma}_k$ $(k=1,\ldots,n)$. Then equations (4.22) with $m=n+1$ imply

$$\mathbf{S}'\mathbf{G}\mathbf{S} = \boldsymbol{\Lambda}, \tag{4.37}$$

where $\boldsymbol{\Lambda}$ is a diagonal matrix with elements $\boldsymbol{\sigma}_k{}'\mathbf{G}\boldsymbol{\sigma}_k$ $(k=1,\ldots,n)$. Equation (4.37) gives

$$\begin{aligned}
\mathbf{G}^{-1} &= \mathbf{S}\boldsymbol{\Lambda}^{-1}\mathbf{S}' \\
&= \sum_{k=1}^{n} (\boldsymbol{\Lambda}^{-1})_{kk}\boldsymbol{\sigma}_k\boldsymbol{\sigma}_k{}' \\
&= \sum_{k=1}^{n} (\boldsymbol{\sigma}_k{}'\mathbf{G}\boldsymbol{\sigma}_k)^{-1}\boldsymbol{\sigma}_k\boldsymbol{\sigma}_k{}' \\
&= \sum_{k=1}^{n} (\boldsymbol{\sigma}_k\boldsymbol{\sigma}_k{}'/\boldsymbol{\sigma}_k{}'\boldsymbol{\gamma}_k), \qquad \text{using equation (4.20)}, \\
&= \sum_{k=1}^{n} \mathbf{A}_k,
\end{aligned}$$

from equation (4.18).

The matrices $\mathbf{B}_k$ are chosen so that the vectors $\boldsymbol{\sigma}_k$ satisfy the eigenvalue properties of equations (4.23). We therefore require

$$\mathbf{B}_k\boldsymbol{\gamma}_k = -\mathbf{H}_k\boldsymbol{\gamma}_k, \qquad \text{cf. equation (4.21)},$$

and

$$\mathbf{B}_m\boldsymbol{\gamma}_k = 0 \quad (k<m), \qquad \text{cf. equations (4.33) } \textit{et seq.}$$

These equations are satisfied if we choose

$$\mathbf{B}_k = -\mathbf{H}_k\boldsymbol{\gamma}_k\mathbf{z}'/\mathbf{z}'\boldsymbol{\gamma}_k,$$

where $\mathbf{z}'$ is a row vector determined by the condition that $\mathbf{B}_k$ is symmetric. Thus

$$\mathbf{B}_k = -\mathbf{H}_k\boldsymbol{\gamma}_k\boldsymbol{\gamma}_k{}'\mathbf{H}_k/\boldsymbol{\gamma}_k{}'\mathbf{H}_k\boldsymbol{\gamma}_k,$$

and it follows from equation (4.17) that $\mathbf{H}_k$ is symmetric.

It should be noted that $\mathbf{A}_k$ and $\mathbf{B}_k$ are $(n \times n)$ symmetric matrices of rank 1, and that $\mathbf{A}_k + \mathbf{B}_k$ is of rank 2. It follows that $\mathbf{H}_k$ is updated by the addition of a symmetric matrix of rank 2. Quasi-Newton methods are often classified in terms of the matrix used to update $\mathbf{H}_k$; thus the DFP method is a 'rank 2' method. In general, methods of higher rank have better convergence properties than those of lower rank, but this gain is offset by the greater computational effort required in each iteration.

The following example illustrates the application of the DFP method to the problem of minimizing a simple quadratic function.

Example 4.2

Solve by the DFP method:

$$\text{minimize } f(\mathbf{x}) \equiv x_1^2 - x_1 x_2 + 3x_2^2,$$

starting at the point $\mathbf{x}_1 = [1,2]$.

[*Note.* This example is used throughout the rest of the chapter to illustrate the various gradient methods.]

Solution

The gradient vector is

$$\mathbf{g}(\mathbf{x}) \equiv \nabla f(\mathbf{x}) \equiv [2x_1 - x_2, -x_1 + 6x_2].$$

We find, in turn,

$$\mathbf{g}_1 = [0,11], \ \mathbf{d}_1 = -\mathbf{H}_1\mathbf{g}_1 = -\mathbf{g}_1 = [0,-11],$$

$$\lambda_1{}^* = 1/6, \ \boldsymbol{\sigma}_1 = [0,-11/6];$$

$$\mathbf{x}_2 = [1,1/6], \ \mathbf{g}_2 = [11/6,0], \ \boldsymbol{\gamma}_1 = \mathbf{g}_2 - \mathbf{g}_1 = [11/6,-11],$$

$$\mathbf{A}_1 = \begin{pmatrix} 0 & 0 \\ 0 & 1/6 \end{pmatrix} \quad \mathbf{B}_1 = \begin{pmatrix} -1/37 & 6/37 \\ 6/37 & -36/37 \end{pmatrix},$$

$$\mathbf{H}_2 = \begin{pmatrix} 36/37 & 6/37 \\ 6/37 & 43/222 \end{pmatrix}, \quad \mathbf{d}_2 = [-66/37, -11/37],$$

$$\lambda_2{}^* = 37/66, \ \boldsymbol{\sigma}_2 = [-1,-1/6].$$

Then $\mathbf{x}_3 = [0,0]$, $\mathbf{g}_3 = [0,0]$. Hence $\mathbf{x}_3$ is the optimal point, i.e.

$$\mathbf{x}^* = [0,0]; \ f(\mathbf{x}^*) = 0.$$

The method has terminated, as expected, in two iterations.

To check that $\mathbf{H}_3 = \mathbf{G}^{-1}$ (cf. Theorem 4.3), we require

$$\mathbf{A}_2 = \begin{pmatrix} 6/11 & 1/11 \\ 1/11 & 1/66 \end{pmatrix}, \quad \mathbf{B}_2 = \begin{pmatrix} -36/37 & -6/37 \\ -6/37 & -1/37 \end{pmatrix}, \quad \mathbf{G} = \begin{pmatrix} 2 & -1 \\ -1 & 6 \end{pmatrix}.$$

Then

$$\mathbf{H}_3 = \mathbf{H}_2 + \mathbf{A}_2 + \mathbf{B}_2 = \begin{pmatrix} 6/11 & 1/11 \\ 1/11 & 2/11 \end{pmatrix} = \mathbf{G}^{-1}.$$

As a further check on the theory—cf. equation (4.36)—note that

$$\mathbf{A}_1 + \mathbf{A}_2 = \begin{pmatrix} 6/11 & 1/11 \\ 1/11 & 2/11 \end{pmatrix} = \mathbf{G}^{-1}.$$

4.5 THE COMPLEMENTARY DFP FORMULA

The central formula of the Davidon–Fletcher–Powell method is the updating formula for $\mathbf{H}_k$, equation (4.17). An alternative formula for updating $\mathbf{H}_k$, which appears to be superior to equation (4.17), was discovered independently and in different ways by Broyden[15] and Fletcher.[34] The following analysis follows that of Fletcher.

Let $\boldsymbol{\Gamma}_k = \mathbf{H}_k^{-1}$. Then equation (4.17) can be written

$$\boldsymbol{\Gamma}_{k+1} = \left(\mathbf{I} - \frac{\gamma_k\sigma_k'}{\sigma_k'\gamma_k}\right)\boldsymbol{\Gamma}_k\left(\mathbf{I} - \frac{\sigma_k\gamma_k'}{\sigma_k'\gamma_k}\right) + \frac{\gamma_k\gamma_k'}{\sigma_k'\gamma_k}. \tag{4.38}$$

For, with this expression for $\boldsymbol{\Gamma}_{k+1}$, we find that

$$\begin{aligned}\mathbf{H}_{k+1}\boldsymbol{\Gamma}_{k+1} &= \mathbf{H}_{k+1}\left(\mathbf{H}_k^{-1} - \frac{\mathbf{H}_k^{-1}\sigma_k\gamma_k'}{\sigma_k'\gamma_k} - \frac{\gamma_k\sigma_k'\mathbf{H}_k^{-1}}{\sigma_k'\gamma_k} + \frac{\gamma_k\sigma_k'\mathbf{H}_k^{-1}\sigma_k\gamma_k'}{(\sigma_k'\gamma_k)^2}\right.\\ &\qquad\left. + \frac{\gamma_k\gamma_k'}{\sigma_k'\gamma_k}\right)\\ &= (\mathbf{H}_k + \mathbf{A}_k + \mathbf{B}_k)\left(\mathbf{H}_k^{-1} - \frac{\mathbf{H}_k^{-1}\sigma_k\gamma_k'}{\sigma_k'\gamma_k}\right) - \frac{\sigma_k\sigma_k'\mathbf{H}_k^{-1}}{\sigma_k'\gamma_k}\\ &\qquad + \frac{\sigma_k\sigma_k'\mathbf{H}_k^{-1}\sigma_k\gamma_k'}{(\sigma_k'\gamma_k)^2} + \frac{\sigma_k\gamma_k'}{\sigma_k'\gamma_k},\end{aligned}$$

using the result from equation (4.21) that

$$\mathbf{H}_{k+1}\gamma_k = \sigma_k. \tag{4.39}$$

Substituting for $\mathbf{A}_k$ and $\mathbf{B}_k$ from equations (4.18) and (4.19), we find after a little manipulation that

$$\mathbf{H}_{k+1}\boldsymbol{\Gamma}_{k+1} = \mathbf{I},$$

i.e.

$$\boldsymbol{\Gamma}_{k+1} = \mathbf{H}_{k+1}^{-1},$$

which verifies equation (4.38).

From equation (4.39), we have

$$\boldsymbol{\Gamma}_{k+1}\sigma_k = \gamma_k. \tag{4.40}$$

Comparing equations (4.39) and (4.40) and remembering that equation (4.38) is merely the DFP formula in a new notation, we see that a new formula for updating $\mathbf{H}_k$ may be obtained from equation (4.38) by interchanging σ_k and γ_k and replacing $\boldsymbol{\Gamma}_{k+1}, \boldsymbol{\Gamma}_k$ by $\mathbf{H}_{k+1}, \mathbf{H}_k$, respectively. This gives the *complementary DFP formula*:

$$\begin{aligned}\mathbf{H}_{k+1} &= \left(\mathbf{I} - \frac{\sigma_k\gamma_k'}{\gamma_k'\sigma_k}\right)\mathbf{H}_k\left(\mathbf{I} - \frac{\gamma_k\sigma_k'}{\gamma_k'\sigma_k}\right) + \frac{\sigma_k\sigma_k'}{\gamma_k'\sigma_k}\\ &= \mathbf{H}_k + \frac{1}{\gamma_k'\sigma_k}(\beta_k\sigma_k\sigma_k' - \sigma_k\gamma_k'\mathbf{H}_k - \mathbf{H}_k\gamma_k\sigma_k'),\end{aligned} \tag{4.41}$$

where

$$\beta_k = 1 + \frac{\gamma_k'\mathbf{H}_k\gamma_k}{\gamma_k'\sigma_k}.$$

This is a rank 2 formula, since the columns of the updating matrix belong to the space spanned by the vectors σ_k and $\mathbf{H}_k\gamma_k$.

Since equation (4.41) for updating $\mathbf{H}_k$ was obtained from equation (4.38) by a mere change of notation and since equation (4.38) is equivalent to equation (4.17), it follows that the corresponding formula for updating $\mathbf{\Gamma}_k$ may be obtained from equation (4.17) by the inverse change of notation. This gives

$$\mathbf{\Gamma}_{k+1} = \mathbf{\Gamma}_k + \frac{\gamma_k\gamma_k'}{\gamma_k'\sigma_k} - \frac{\mathbf{\Gamma}_k\sigma_k\sigma_k'\mathbf{\Gamma}_k}{\sigma_k'\mathbf{\Gamma}_k\sigma_k}. \tag{4.42}$$

The pair of formulae (4.41) and (4.42) may be regarded as dual to the pair (4.17) and (4.38), in the sense that each pair may be derived from the other by the interchange

$$(\mathbf{H}_k, \sigma_k) \leftrightarrow (\mathbf{\Gamma}_k, \gamma_k).$$

Finally, we discuss without proof some properties of the complementary DFP formula and of the minimization algorithm based on it. Broyden[15] shows that the algorithm possesses all the desirable properties of the DFP method and, in addition, a property that the DFP method does not possess, namely, that a certain matrix error norm is reduced strictly monotonically when a quadratic function is being minimized. Numerically, the complementary DFP formula avoids the tendency which is present in the DFP method for the matrices $\mathbf{H}_k$ to become singular.

Fletcher[34] shows that equation (4.41) can be used successfully in a minimization algorithm without the need for a linear search, provided that suitable step lengths are chosen. He suggests the formula

$$\sigma_k = -\lambda_k\mathbf{H}_k\mathbf{g}_k,$$

where λ_k is chosen so that the change in function value Δf_k satisfies

$$\frac{\Delta f_k}{\mathbf{g}_k'\sigma_k} \geqslant \mu \qquad (0 < \mu \ll 1).$$

A value of λ_k satisfying this criterion can be found by trying

$$\lambda_k = 1, w, w^2, \ldots \qquad (0 < w < 1)$$

in succession. In practice, the values $\mu = 0{\cdot}0001$ and $w = 0{\cdot}1$ have been used. It is essential, however, that the matrices $\mathbf{H}_k$ remain positive definite. For quadratic functions, this is a property of equation (4.41); for non-quadratic functions, the property is ensured by imposing the extra condition $\sigma_k'\gamma_k > 0$. A value of σ_k which satisfies this condition can always be found by altering the above rules to allow λ_k to become greater than unity, if necessary.

Needless to say, important savings in computational effort are achieved by dispensing with linear searches. A theoretical reason for using linear searches is that the algorithm possesses quadratic termination (page 113) when the searches are made exactly. However, even when the linear searches are abandoned, the algorithm retains the important properties that the matrices $\mathbf{H}_k$ remain positive definite and that $\mathbf{H}_k$ tends to the inverse Hessian matrix.

Powell[65,66] has examined the quadratic termination properties of many gradient algorithms, including this one, and in particular the role played by linear searches. His results lead to a new (1972) algorithm for unconstrained minimization without linear searches.

Broyden[14] discusses the general theoretical background of the updating formula (4.41). For further discussion of the complementary DFP formula, the reader is referred to the original papers by Broyden[15] and Fletcher;[34] the latter is concerned only with the properties of the method in the absence of linear searches.

Exercise

Solve the problem of Example 4.2 using the complementary DFP formula.

4.6 CONJUGATE DIRECTIONS

Consider the problem of finding the minimum value of the quadratic function

$$f(\mathbf{x}) \equiv \tfrac{1}{2}\mathbf{x}'\mathbf{G}\mathbf{x} + \mathbf{b}'\mathbf{x} + c, \tag{4.43}$$

where $\mathbf{G}$ is positive definite and symmetric. In two dimensions, the level curves $f(\mathbf{x}) = K$, for different values of K, are concentric ellipses (Figure 4.1). Suppose

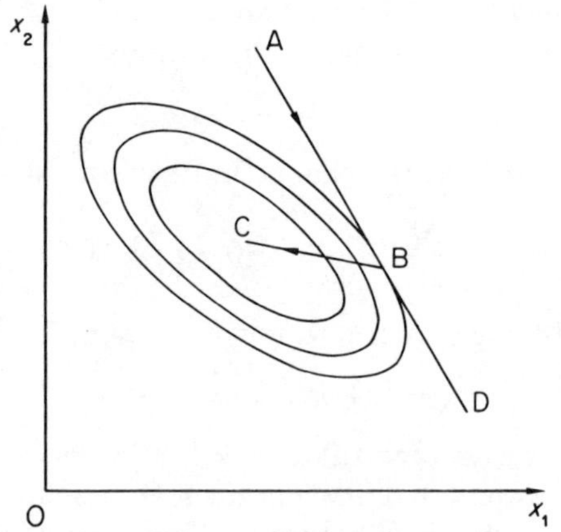

Figure 4.1 Conjugate directions

that we search for a minimum from the point A in the direction AD, that this minimum occurs at B and that C is the optimal point. We say that the direction BC is *conjugate* to the direction AD since, for any ellipse $f(\mathbf{x}) = K$, the diameter through B is conjugate (in the usual geometrical sense) to the diameter parallel to AD. The idea of conjugate directions is easily extended to n dimensions by means of the following definition.

Let $\mathbf{G}$ be a positive definite symmetric matrix. Then the directions represented by the vectors $\mathbf{p} \neq \mathbf{0}$ and $\mathbf{q} \neq \mathbf{0}$ are said to be *conjugate* with respect to $\mathbf{G}$ if $\mathbf{p}'\mathbf{G}\mathbf{q} = 0$. (The phrase 'with respect to $\mathbf{G}$' is often omitted when there is no ambiguity.)

Several methods are available for generating sets of mutually conjugate directions; any such method may be regarded as the major part of an algorithm for the minimization of $f(\mathbf{x})$. We have already noted (page 115) that the DFP method generates a set of mutually conjugate directions. The more important properties of conjugate directions are established in the following three theorems.

***Theorem* 4.4**

If the vectors $\mathbf{d}_j$ are mutually conjugate, then they are linearly independent.

Proof

Suppose that

$$\sum_j \lambda_j \mathbf{d}_j = \mathbf{0}$$

for some scalars λ_j. If k is *any* one of the values of j, then

$$\mathbf{d}_k'\mathbf{G} \sum_j \lambda_j \mathbf{d}_j = 0,$$

i.e.

$$\lambda_k \mathbf{d}_k'\mathbf{G}\mathbf{d}_k = 0,$$

all the remaining terms vanishing because of the mutual conjugacy. Now $\mathbf{G}$ is positive definite and $\mathbf{d}_k \neq \mathbf{0}$; hence $\lambda_k = 0$. It follows that the $\mathbf{d}_j$ are linearly independent.

***Theorem* 4.5**

Let $\mathbf{x}_k$ and $\mathbf{x}_{k+1}$ be consecutive current points in a minimization algorithm when the objective function is the quadratic $f(\mathbf{x})$ of equation (4.43). *If*

(i) $\mathbf{x}_k$ *minimizes $f(\mathbf{x})$ in the direction $\mathbf{d}_l$,*
(ii) $\mathbf{x}_{k+1}$ *minimizes $f(\mathbf{x})$ in the direction $\mathbf{d}_m$,*
(iii) $\mathbf{d}_l$ *and $\mathbf{d}_m$ are conjugate directions,*

then $\mathbf{x}_{k+1}$ also minimizes $f(\mathbf{x})$ in the direction $\mathbf{d}_l$.

Proof

Condition (i) implies

$$\mathbf{d}_l'\mathbf{g}_k = 0. \tag{4.44}$$

Also, from (iii),

$$\mathbf{d}_l'\mathbf{G}\mathbf{d}_m = 0. \tag{4.45}$$

For the quadratic function (4.43), we have

$$\mathbf{g}_{k+1} - \mathbf{g}_k = \mathbf{G}(\mathbf{x}_{k+1} - \mathbf{x}_k), \tag{4.46}$$

and (ii) implies

$$\mathbf{x}_{k+1} = \mathbf{x}_k + \lambda_m{}^* \mathbf{d}_m, \tag{4.47}$$

where $\lambda_m{}^*$ is the value of λ_m which minimizes $f(\mathbf{x}_k + \lambda_m \mathbf{d}_m)$. Hence, from equations (4.44) to (4.47),

$$\mathbf{d}_l{}' \mathbf{g}_{k+1} = \mathbf{d}_l{}'(\mathbf{g}_k + \lambda_m{}^* \mathbf{G}\mathbf{d}_m) = 0,$$

which proves the theorem.

[Note that Theorem 4.5 is still true if hypothesis (ii), or equivalently equation (4.47), is replaced by
(ii)′ $\mathbf{x}_{k+1} = \mathbf{x}_k + \lambda_m \mathbf{d}_m$ for *any* value of λ_m.
Hypothesis (ii) is, however, satisfied for appropriate directions $\mathbf{d}_m$ in all the subsequent applications.]

***Corollary* 1**

If

(i) $\mathbf{x}_k$ *minimizes* $f(\mathbf{x})$ *in the directions* $\mathbf{d}_1, \ldots, \mathbf{d}_r$,
(ii) $\mathbf{x}_{k+1}$ *minimizes* $f(\mathbf{x})$ *in the direction* $\mathbf{d}_k$,
(iii) $\mathbf{d}_k$ *is conjugate to the directions* $\mathbf{d}_1, \ldots, \mathbf{d}_r$,

then $\mathbf{x}_{k+1}$ *minimizes* $f(\mathbf{x})$ *in the directions* $\mathbf{d}_1, \ldots, \mathbf{d}_r$ *and* $\mathbf{d}_k$.

Proof

Let $m = k$ in Theorem 4.5. The corollary then follows immediately from the theorem by setting l equal to $1, \ldots, r$, in turn.

***Corollary* 2**

If, for $k \geqslant 2$,

(i) $\mathbf{x}_{r+1}$ *minimizes* $f(\mathbf{x})$ *in the direction* $\mathbf{d}_r$ *for* $r = 1, \ldots, k$,
(ii) *each of the pairs of directions* $(\mathbf{d}_1, \mathbf{d}_2)$, $(\mathbf{d}_2, \mathbf{d}_3)$, ..., $(\mathbf{d}_{k-1}, \mathbf{d}_k)$ *is a conjugate pair,*

then $\mathbf{x}_{r+1}$ *minimizes* $f(\mathbf{x})$ *in the directions* $\mathbf{d}_1, \ldots, \mathbf{d}_r$ *for* $r = 1, \ldots, k$.

Proof

The corollary follows by repeated application of the theorem, with (k, l, m) taking the values (2,1,2); (3,1,2), (3,2,3); (4,1,2), (4,2,3), (4,3,4); ....; $(k,1,2)$, $(k,2,3)$, ..., $(k, k-1, k)$.

***Theorem* 4.6**

Let $\mathbf{d}_i$, $i = 1, \ldots, m$ $(\leqslant n)$, *be mutually conjugate directions. Then the global minimum of* $f(\mathbf{x})$ *of equation* (4.43), *in the subspace* E^m *containing the point* $\mathbf{x}_1$ *and the directions* $\mathbf{d}_i$, *may be found by searching along each of these directions once only.*

Proof

The required minimum occurs at the point

$$\mathbf{x}_1 + \sum_i \lambda_i \mathbf{d}_i,$$

where the parameters λ_i are chosen so as to minimize

$$f(\mathbf{x}_1 + \sum_i \lambda_i \mathbf{d}_i) = \tfrac{1}{2}\sum_i \lambda_i{}^2 \mathbf{d}_i{}'\mathbf{G}\mathbf{d}_i + \sum_i \lambda_i \mathbf{d}_i{}'(\mathbf{G}\mathbf{x}_1 + \mathbf{b}) + f(\mathbf{x}_1). \qquad (4.48)$$

There is no term in $\lambda_r\lambda_s$ $(r \neq s)$ on the right-hand side because the directions $\mathbf{d}_i$ are mutually conjugate. Hence the search in the direction $\mathbf{d}_r$ finds the value of λ_r which minimizes

$$\tfrac{1}{2}\lambda_r{}^2 \mathbf{d}_r{}'\mathbf{G}\mathbf{d}_r + \lambda_r \mathbf{d}_r{}'(\mathbf{G}\mathbf{x}_1 + \mathbf{b}),$$

and this value of λ_r is independent of all the other terms on the right-hand side of equation (4.48). In other words, the global minimum of $f(\mathbf{x})$ in the m-dimensional subspace defined in the theorem may be found by searching along each of the directions $\mathbf{d}_i$ once only.

Corollary

An algorithm that uses mutually conjugate directions of search possesses the property of quadratic termination.

4.7 THE FLETCHER–REEVES METHOD

In 1964, Fletcher and Reeves[37] derived a simple recurrence formula which produces a sequence of mutually conjugate directions. We shall first derive this formula and then go on to discuss some computational aspects of the associated minimization algorithm.

Consider the problem of minimizing the quadratic function

$$f(\mathbf{x}) \equiv \tfrac{1}{2}\mathbf{x}'\mathbf{G}\mathbf{x} + \mathbf{b}'\mathbf{x} + c, \qquad (4.49)$$

where $\mathbf{G}$ is positive definite and symmetric, by successive linear searches along mutually conjugate directions. The initial step is in the direction of steepest descent:

$$\mathbf{d}_1 = -\mathbf{g}_1.$$

Subsequently, the mutually conjugate directions are chosen so that

$$\mathbf{d}_{k+1} = -\mathbf{g}_{k+1} + \sum_{r=1}^{k} \beta_r \mathbf{d}_r \quad (k=1,2,\ldots.),$$

where the coefficients β_r are to be determined. It turns out that all the coefficients except β_k are zero, giving

$$\mathbf{d}_{k+1} = -\mathbf{g}_{k+1} + \beta_k \mathbf{d}_k \quad (k=1,2,\ldots.), \qquad (4.50)$$

where

$$\beta_k = \mathbf{g}_{k+1}{}^2/\mathbf{g}_k{}^2. \tag{4.51}$$

We shall now prove that equations (4.50) and (4.51) produce mutually conjugate directions. We show first that equations (4.50) imply that for $k=1,2,....$, the space spanned by $\mathbf{d}_1,....,\mathbf{d}_k$ is precisely the space spanned by $\mathbf{g}_1, \mathbf{G}\mathbf{g}_1,....,$ $\mathbf{G}^{k-1}\mathbf{g}_1$.

For the quadratic function (4.49), we have

$$\mathbf{g}_{k+1} = \mathbf{g}_k + \lambda_k{}^*\mathbf{G}\mathbf{d}_k \qquad (k=1,2,....), \tag{4.52}$$

where $\lambda_k{}^*$ is the value of λ_k which minimizes $f(\mathbf{x}_k + \lambda_k \mathbf{d}_k)$. Let $L(\mathbf{a},\mathbf{b},\mathbf{c},....)$ denote a linear combination of the vectors $\mathbf{a},\mathbf{b},\mathbf{c},.....$ Then, using equations (4.50) and (4.52), we have

$$\begin{aligned}
\mathbf{d}_1 &= -\mathbf{g}_1;\\
\mathbf{d}_2 &= -\mathbf{g}_2 + \beta_1\mathbf{d}_1\\
&= L(\mathbf{g}_1, \mathbf{G}\mathbf{d}_1, \mathbf{d}_1)\\
&= L(\mathbf{g}_1, \mathbf{G}\mathbf{g}_1);\\
\mathbf{d}_3 &= -\mathbf{g}_3 + \beta_2\mathbf{d}_2\\
&= L(\mathbf{g}_2, \mathbf{G}\mathbf{d}_2, \mathbf{d}_2)\\
&= L(\mathbf{g}_1, \mathbf{G}\mathbf{g}_1, \mathbf{G}^2\mathbf{g}_1);
\end{aligned}$$

and so on, up to

$$\mathbf{d}_k = L(\mathbf{g}_1, \mathbf{G}\mathbf{g}_1, ..., \mathbf{G}^{k-1}\mathbf{g}_1). \tag{4.53}$$

The main part of the proof is by mathematical induction. Assume that equations (4.50) produce mutually conjugate directions $\mathbf{d}_1,....,\mathbf{d}_k$. We shall prove that $\mathbf{d}_1,....,\mathbf{d}_{k+1}$ are mutually conjugate.

The point $\mathbf{x}_{k+1}$ is given by

$$\mathbf{x}_{k+1} = \mathbf{x}_k + \lambda_k{}^*\mathbf{d}_k,$$

and therefore $\mathbf{x}_{k+1}$ minimizes $f(\mathbf{x})$ in the directions $\mathbf{d}_1,....,\mathbf{d}_k$, by Corollary 1 of Theorem 4.5 with $r=k-1$. Hence $\mathbf{g}_{k+1}$ is orthogonal to $\mathbf{d}_1,....,\mathbf{d}_k$ and, by equation (4.53) and the equations immediately preceding it, is also orthogonal to $\mathbf{g}_1, \mathbf{G}\mathbf{g}_1,....,\mathbf{G}^{k-1}\mathbf{g}_1$. However, for any value of r,

$$\mathbf{g}_{k+1}{}'(\mathbf{G}^r\mathbf{g}_1) = 0 \Rightarrow \mathbf{g}_{k+1}{}'\mathbf{G}(\mathbf{G}^{r-1}\mathbf{g}_1) = 0.$$

Hence $\mathbf{g}_{k+1}$ is conjugate to every vector in the space spanned by $\mathbf{g}_1, \mathbf{G}\mathbf{g}_1,....,$ $\mathbf{G}^{k-2}\mathbf{g}_1$ and is therefore conjugate to the directions $\mathbf{d}_1,....,\mathbf{d}_{k-1}$ which are contained in this space. Thus $-\mathbf{g}_{k+1}$, as well as $\mathbf{d}_k$, is already conjugate to the directions $\mathbf{d}_1,....,\mathbf{d}_{k-1}$, and so the addition of an appropriate multiple of $\mathbf{d}_k$ to $-\mathbf{g}_{k+1}$ provides the required direction $\mathbf{d}_{k+1}$; this explains the form of equation (4.50).

The coefficient β_k in equation (4.50) is determined by the condition

$$\mathbf{d}_{k+1}{}'\mathbf{G}\mathbf{d}_k = 0.$$

Using equation (4.50), this gives

$$-\mathbf{g}_{k+1}'\mathbf{G}\mathbf{d}_k + \beta_k \mathbf{d}_k'\mathbf{G}\mathbf{d}_k = 0.$$

Hence

$$\begin{aligned}\beta_k &= \mathbf{g}_{k+1}'\mathbf{G}\mathbf{d}_k/\mathbf{d}_k'\mathbf{G}\mathbf{d}_k \\ &= \mathbf{g}_{k+1}'(\mathbf{g}_{k+1} - \mathbf{g}_k)/\lambda_k^*(-\mathbf{g}_k' + \beta_{k-1}\mathbf{d}_{k-1}')\mathbf{G}\mathbf{d}_k,\end{aligned}$$

from equations (4.52) and (4.50),

$$= \mathbf{g}_{k+1}'(\mathbf{g}_{k+1} - \mathbf{g}_k)/\lambda_k^*(-\mathbf{g}_k'\mathbf{G}\mathbf{d}_k). \tag{4.54}$$

Now $\mathbf{g}_{k+1}$ is orthogonal to $\mathbf{d}_1,....,\mathbf{d}_k$, as stated above, and $\mathbf{g}_k = L(\mathbf{d}_{k-1},\mathbf{d}_k)$, from equation (4.50). It follows that $\mathbf{g}_{k+1}'\mathbf{g}_k = 0$. Also,

$$\lambda_k^*(-\mathbf{g}_k'\mathbf{G}\mathbf{d}_k) = -\mathbf{g}_k'(\mathbf{g}_{k+1} - \mathbf{g}_k) = \mathbf{g}_k^2,$$

and hence, from equation (4.54),

$$\beta_k = \mathbf{g}_{k+1}^2/\mathbf{g}_k^2 \qquad (k=1,2,....),$$

which verifies equation (4.51).

We have now proved that if equation (4.50) produces mutually conjugate directions $\mathbf{d}_1,....,\mathbf{d}_k$, then it produces mutually conjugate directions $\mathbf{d}_1,....,\mathbf{d}_{k+1}$. It only remains to prove that $\mathbf{d}_1$ and $\mathbf{d}_2$ are conjugate directions. From equations (4.50), (4.51) and (4.52), with $k = 1$, we find

$$\begin{aligned}\mathbf{d}_2'\mathbf{G}\mathbf{d}_1 &= \left(-\mathbf{g}_2' - \frac{\mathbf{g}_2^2}{\mathbf{g}_1^2}\mathbf{g}_1'\right)\left(\frac{\mathbf{g}_2 - \mathbf{g}_1}{\lambda_1^*}\right) \\ &= \frac{1}{\lambda_1^*}\left(-\mathbf{g}_2^2 + \frac{\mathbf{g}_2^2}{\mathbf{g}_1^2}\mathbf{g}_1^2\right) \\ &= 0,\end{aligned}$$

where we have used the relations

$$\mathbf{g}_1'\mathbf{g}_2 = \mathbf{g}_2'\mathbf{g}_1 = -\mathbf{g}_2'\mathbf{d}_1 = 0.$$

Hence $\mathbf{d}_1$ and $\mathbf{d}_2$ are conjugate directions and the proof of equations (4.50) and (4.51) is complete.

Theorem 4.6 shows that the Fletcher–Reeves method will find the minimum of a positive definite quadratic function of n variables in, at most, n iterations. When used on non-quadratic functions the method is iterative. Fletcher and Reeves suggest that the direction of search should revert periodically to the direction of steepest descent, all previous information on directions being discarded. With this procedure, the algorithm retains the property of quadratic termination provided that such re-starts are not made more often than every nth iteration; satisfactory results are obtained if the direction of steepest descent is used for $\mathbf{d}_1, \mathbf{d}_{n+1}, \mathbf{d}_{2n+1}, ...$.

The cubic interpolation method of Davidon (Section 3.9) is recommended for the linear searches, with $\kappa = |\mathbf{d}_k|^{-1}$ in equation (3.23), where it is again assumed that $G_0 < 0$. If $G_0 > 0$, then equation (3.23) is replaced by $\alpha = \kappa = |\mathbf{d}_k|^{-1}$. The value $|\mathbf{d}_k|^{-1}$ for κ is arbitrary, though convenient. Fletcher and Reeves use the fairly stringent convergence criterion of stopping only when a complete cycle of $(n+1)$ iterations, beginning with a steepest descent search, produces no further reduction in the value of the objective function.

Unlike the DFP method, which produces conjugate directions by means of *matrix* formulae, the Fletcher–Reeves method uses only *vector* formulae. This difference becomes significant when n is so large that problems of computer storage arise. The amount of computer storage required is of order n, compared with order n^2 for the DFP method.

For further discussion of the properties of conjugate direction methods, the reader should consult the article by Fletcher (Chapter 5, 'Conjugate Direction Methods').[55]

Example 4.3

Solve by the Fletcher–Reeves method:

$$\textit{minimize} \quad f(\mathbf{x}) \equiv x_1{}^2 - x_1 x_2 + 3x_2{}^2,$$

starting at the point $\mathbf{x}_1 = [1,2]$.

Solution

The gradient vector is

$$\mathbf{g}(\mathbf{x}) \equiv \nabla f(\mathbf{x}) \equiv [2x_1 - x_2, -x_1 + 6x_2].$$

Hence the search direction from $\mathbf{x}_1$ is

$$\mathbf{d}_1 = -\mathbf{g}_1 = [0, -11],$$

giving a minimum at $\mathbf{x}_2 = [1, 1/6]$. The search direction from $\mathbf{x}_2$ is

$$\mathbf{d}_2 = -\mathbf{g}_2 + \beta_1 \mathbf{d}_1,$$

where $\mathbf{g}_2 = [11/6, 0]$ and $\beta_1 = -\mathbf{g}_2{}^2/\mathbf{g}_1{}^2 = 1/36$. Therefore $\mathbf{d}_2 = [-11/6, -11/36]$ and the minimum in this direction is at $\mathbf{x}_3 = [0,0]$. Since $\mathbf{g}_3 = [0,0]$, the point $\mathbf{x}_3$ is optimal, i.e.

$$\mathbf{x}^* = [0,0]; \quad f(\mathbf{x}^*) = 0.$$

4.8 SMITH'S METHOD

The gradient methods of Sections 4.2 to 4.7 rely for their effectiveness on the fact that values of the components of the gradient vector $\mathbf{g}(\mathbf{x}) \equiv \nabla f(\mathbf{x})$ are readily available. It sometimes happens, however, that the evaluation of these components requires a prohibitive amount of computational effort. In such cases, it is still possible to use the powerful technique of linear searches along mutually conjugate directions—any linear search technique that does not involve the evaluation of derivatives is appropriate, e.g. Powell's quadratic interpolation method (Section 3.8). The outstanding problem is to construct a sequence of mutually conjugate directions without using the gradient vector explicitly.

In this section we shall describe Smith's method,[72] which is the simplest method for producing conjugate directions without calculating derivatives. In the next section, we consider two methods due to Powell which were developed directly from Smith's method.

Given the quadratic objective function

$$f(\mathbf{x}) \equiv \tfrac{1}{2}\mathbf{x}'\mathbf{G}\mathbf{x} + \mathbf{b}'\mathbf{x} + c, \tag{4.55}$$

where $\mathbf{G}$ is positive definite and symmetric, we obtain a direction conjugate

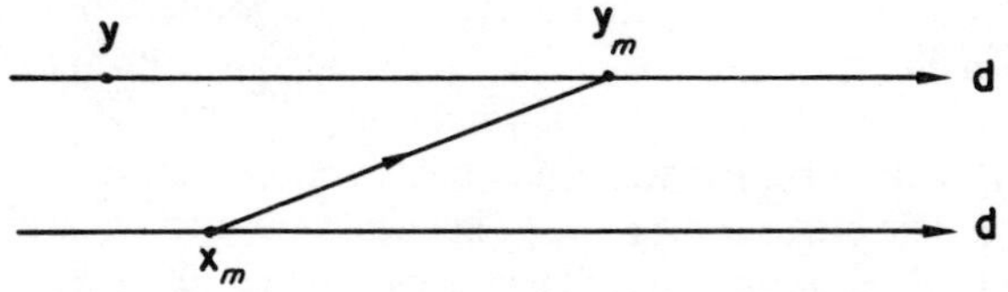

Figure 4.2 Construction of a direction conjugate to a given direction

to $\mathbf{d}$ as follows. Let $\mathbf{x}_m$ minimize $f(\mathbf{x})$ in the direction $\mathbf{d}$ (Figure 4.2) and let $\mathbf{y}$ be a point that is not on the line $\mathbf{r} = \mathbf{x}_m + s\mathbf{d}$. If $\mathbf{y}_m$ minimizes $f(\mathbf{x})$ on the line $\mathbf{r} = \mathbf{y} + t\mathbf{d}$, then the direction $\mathbf{y}_m - \mathbf{x}_m$ is conjugate to $\mathbf{d}$. For

$$\mathbf{g}(\mathbf{x}_m) = \mathbf{G}\mathbf{x}_m + \mathbf{b} \quad \text{and} \quad \mathbf{g}(\mathbf{y}_m) = \mathbf{G}\mathbf{y}_m + \mathbf{b},$$

from equation (4.55). Therefore

$$\mathbf{d}'(\mathbf{G}\mathbf{x}_m + \mathbf{b}) = 0 \quad \text{and} \quad \mathbf{d}'(\mathbf{G}\mathbf{y}_m + \mathbf{b}) = 0.$$

By subtraction,

$$\mathbf{d}'\mathbf{G}(\mathbf{y}_m - \mathbf{x}_m) = 0,$$

as required.

Each iteration of Smith's method is divided into n stages. The number of linear searches in the kth stage is k, so that the total number of linear searches in one iteration is $\tfrac{1}{2}n(n+1)$. In any iteration, each stage after the first produces a new direction which is conjugate to all the previous directions of that iteration. The method begins with a set of n linearly independent directions, usually the coordinate directions $\mathbf{e}_j$, and the first iteration proceeds as follows.

Stage 1. Search from the initial point $\mathbf{x}_1$ along $\mathbf{d}_1 = \mathbf{e}_1$ for a minimum at $\mathbf{x}_2$.

Stage 2. Let $\mathbf{y}_{20} = \mathbf{x}_2 + t_2\mathbf{e}_2$, where t_2 $(\neq 0)$ is chosen arbitrarily. Search from $\mathbf{y}_{20}$ along $\mathbf{d}_1$ for a minimum at $\mathbf{y}_{21}$. Then search from $\mathbf{y}_{21}$ along $\mathbf{d}_2 = \mathbf{y}_{21} - \mathbf{x}_2$ for a minimum at $\mathbf{x}_3$.

By construction, $\mathbf{d}_2$ is conjugate to $\mathbf{d}_1$ and hence $\mathbf{x}_3$ minimizes $f(\mathbf{x})$ in the directions $\mathbf{d}_1$ and $\mathbf{d}_2$, by Theorem 4.5.

Stage 3. Let $\mathbf{y}_{30} = \mathbf{x}_3 + t_3\mathbf{e}_3$, where t_3 ($\neq 0$) is chosen arbitrarily. Search from $\mathbf{y}_{30}$ along $\mathbf{d}_1$ for a minimum at $\mathbf{y}_{31}$ and from $\mathbf{y}_{31}$ along $\mathbf{d}_2$ for a minimum at $\mathbf{y}_{32}$. Then search from $\mathbf{y}_{32}$ along $\mathbf{d}_3 = \mathbf{y}_{32} - \mathbf{x}_3$ for a minimum at $\mathbf{x}_4$.

By construction, $\mathbf{y}_{32}$ minimizes $f(\mathbf{x})$ in the direction $\mathbf{d}_2$, and $\mathbf{d}_1$ is conjugate to $\mathbf{d}_2$. Therefore $\mathbf{y}_{32}$ minimizes $f(\mathbf{x})$ in the directions $\mathbf{d}_1$ and $\mathbf{d}_2$; but $\mathbf{x}_3$ also minimizes $f(\mathbf{x})$ in the directions $\mathbf{d}_1$ and $\mathbf{d}_2$. Therefore $\mathbf{d}_3 = \mathbf{y}_{32} - \mathbf{x}_3$ is conjugate to the directions $\mathbf{d}_1$ and $\mathbf{d}_2$, and $\mathbf{x}_4$ minimizes $f(\mathbf{x})$ in the directions $\mathbf{d}_1, \mathbf{d}_2$ and $\mathbf{d}_3$.

Stage k. Let $\mathbf{y}_{k0} = \mathbf{x}_k + t_k\mathbf{e}_k$, where t_k ($\neq 0$) is chosen arbitrarily. For $r = 1,2,....,(k-1)$, search from $\mathbf{y}_{k,r-1}$ in the direction $\mathbf{d}_r$ for a minimum at $\mathbf{y}_{kr}$ and let $\mathbf{d}_k = \mathbf{y}_{k,k-1} - \mathbf{x}_k$. Then search from $\mathbf{y}_{k,k-1}$ in the direction $\mathbf{d}_k$ for a minimum at $\mathbf{x}_{k+1}$.

By construction, $\mathbf{d}_k$ is conjugate to $\mathbf{d}_1,....,\mathbf{d}_{k-1}$, because both $\mathbf{y}_{k,k-1}$ and $\mathbf{x}_k$ minimize $f(\mathbf{x})$ in the directions $\mathbf{d}_1,....,\mathbf{d}_{k-1}$. Hence $\mathbf{x}_{k+1}$ minimizes $f(\mathbf{x})$ in the directions $\mathbf{d}_1,....,\mathbf{d}_k$, by Corollary 1 of Theorem 4.5 with $r = k-1$.

After stage n of the first iteration has been completed, the second iteration begins at stage 1. The current point $\mathbf{x}_{n+1}$ becomes the new $\mathbf{x}_1$ and the directions $\mathbf{d}_j$ replace the old directions $\mathbf{e}_j$; similarly for subsequent iterations.

Since one iteration produces n mutually conjugate directions, it follows that the method finds the minimum of a positive definite quadratic function in one iteration, i.e. after $\frac{1}{2}n(n+1)$ linear searches. For non-quadratic functions the method is iterative; in this case, however, it has been found that the convergence is often unsatisfactory unless a close starting approximation is used.

A defect of Smith's method which is largely responsible for its unsatisfactory convergence properties is that the pattern of linear searches is unbalanced. To take the extreme case, $\frac{1}{2}n(n-1)$ linear searches are made before the direction $\mathbf{e}_n$ is used at all. This defect is overcome in Powell's method, which is the subject of the next section.

Example 4.4

Solve by Smith's method:

$$\textit{minimize} \quad f(\mathbf{x}) \equiv x_1^2 - x_1x_2 + 3x_2^2,$$

starting at the point $\mathbf{x}_1 = [1,2]$.

Solution

Stage 1. Search from $\mathbf{x}_1$ in the direction $\mathbf{d}_1 = \mathbf{e}_1$ for a minimum at $\mathbf{x}_2$, which turns out to be the point $\mathbf{x}_1$ itself.

Stage 2. Take $t_2 = 1$. Then

$$\mathbf{y}_{20} = \mathbf{x}_2 + t_2\mathbf{e}_2 = [1,3].$$

Search from $\mathbf{y}_{20}$ in the direction $\mathbf{d}_1$ for a minimum at $\mathbf{y}_{21} = [3/2,3]$. Then search from $\mathbf{y}_{21}$ in the direction $\mathbf{d}_2 = \mathbf{y}_{21} - \mathbf{x}_2 = [1/2,1]$ for a minimum at $\mathbf{x}_3 = [0,0]$.

This completes the first iteration. Since $f(\mathbf{x})$ is a quadratic function and the linear searches have been performed exactly, we have reached the optimal point

$$\mathbf{x}^* = \mathbf{x}_3 = [0,0]; \quad f(\mathbf{x}^*) = 0.$$

4.9 POWELL'S METHOD

Powell's method[62] is a refinement of Smith's method. We begin by describing a simplified form of the method, which we shall refer to as *Powell's basic method*. A modification to Powell's basic method then gives the algorithm known as Powell's method.

Powell's basic method

The kth iteration of this method starts with a current point $\mathbf{x}_k$ and n directions $\mathbf{d}_{kj}$. The $(k+1)$th iteration starts with a new current point $\mathbf{x}_{k+1}$ and a new set of directions as described below; the object is to build up a set of mutually conjugate directions. Each iteration includes only $(n+1)$ linear searches, though, as proved below, n iterations are required to determine the minimum of a positive definite quadratic function. Powell's quadratic interpolation method (Section 3.8) is recommended for the linear searches.

As in Smith's method, the initial directions are usually taken to be the coordinate directions $\mathbf{e}_j$. Thus, at the beginning of the first iteration, $\mathbf{x}_1$ is given and $\mathbf{d}_{1j} = \mathbf{e}_j$. The kth iteration proceeds as follows. Let $\mathbf{y}_{k0} = \mathbf{x}_k$, and for $r = 1, \ldots, n$ search from $\mathbf{y}_{k,r-1}$ in the direction $\mathbf{d}_{kr}$ for a minimum at $\mathbf{y}_{kr}$. Then search from $\mathbf{y}_{kn}$ in the direction $\boldsymbol{\delta}_k = \mathbf{y}_{kn} - \mathbf{x}_k$ for a minimum at $\mathbf{x}_{k+1}$, giving the new current point.

The n directions for the $(k+1)$th iteration are then

$$\mathbf{d}_{k2}, \ldots, \mathbf{d}_{kn}, \boldsymbol{\delta}_k. \tag{4.56}$$

Note that one direction of the kth iteration, namely $\mathbf{d}_{k1}$, has been discarded in favour of a new direction $\boldsymbol{\delta}_k$. Before the $(k+1)$th iteration begins, the respective directions (4.56) are renamed

$$\mathbf{d}_{k+1,1}, \ldots, \mathbf{d}_{k+1,n}.$$

We now prove inductively that Powell's basic method produces n mutually conjugate directions in n iterations. By construction, each of the points $\mathbf{x}_2$ and $\mathbf{y}_{2n}$ minimizes $f(\mathbf{x})$ in the direction $\boldsymbol{\delta}_1 = \mathbf{d}_{2n}$, and so at the end of the second iteration we have a pair of conjugate directions (cf. Figure 4.2) $\boldsymbol{\delta}_1 = \mathbf{d}_{2n}$ and $\boldsymbol{\delta}_2 = \mathbf{y}_{2n} - \mathbf{x}_2$. That is, in the notation for the third iteration, the directions $\mathbf{d}_{3,n-1}$ and $\mathbf{d}_{3n}$ are conjugate; this begins the induction.

Now assume that k iterations have been completed and that the k directions

$$\mathbf{d}_{k+1,n-k+1}, \ldots, \mathbf{d}_{k+1,n} \tag{4.57}$$

are mutually conjugate. These are the last k directions of search in the kth iteration and are also search directions in the $(k+1)$th iteration. It follows from Theorem 4.6 that the starting point $\mathbf{x}_{k+1}$ for the $(k+1)$th iteration minimizes $f(\mathbf{x})$ in a subspace containing these directions, and, again from Theorem 4.6, that the point $\mathbf{y}_{k+1,n}$ defined in the $(k+1)$th iteration also minimizes $f(\mathbf{x})$ in such a subspace. Hence the direction $\boldsymbol{\delta}_{k+1} = \mathbf{y}_{k+1,n} - \mathbf{x}_{k+1}$ is conjugate to the k directions (4.57) (cf. Figure 4.2). Thus, after $(k+1)$ iterations, we have

$(k+1)$ mutually conjugate directions, and the general step of the induction is proved.

In particular, the method produces n mutually conjugate directions in n iterations. By Theorem 4.6, with $m=n$, it therefore finds the minimum of a positive definite quadratic function of n variables in, at most, n iterations.

The method breaks down when the n directions for an iteration become linearly dependent. This happens when the new direction δ_k contains no component of the discarded direction $\mathbf{d}_{k1}$, i.e. when

$$(\mathbf{y}_{kn}-\mathbf{y}_{k0})'(\mathbf{y}_{k1}-\mathbf{y}_{k0})=0.$$

Since the n directions for any iteration belong to the subspace spanned by the n directions of the previous iteration, the subsequent search is restricted to this subspace and the true minimum may not be found.

Example 4.5

Solve by Powell's basic method:

$$\textit{minimize} \quad f(\mathbf{x}) \equiv x_1^2 - x_1x_2 + 3x_2^2,$$

starting at the point $\mathbf{x}_1 = [1,2]$.

Solution

Iteration 1. $\mathbf{x}_1=[1,2]$, $\mathbf{d}_{11}=\mathbf{e}_1$, $\mathbf{d}_{12}=\mathbf{e}_2$.

Search from $\mathbf{y}_{10}=\mathbf{x}_1$ in the direction $\mathbf{d}_{11}=\mathbf{e}_1$ for a minimum at $\mathbf{y}_{11}$, which turns out to be the point $\mathbf{x}_1$ itself. Thus $\mathbf{y}_{11}=\mathbf{y}_{10}$, and, as explained above, the method breaks down: the directions of search in subsequent iterations are restricted to a subspace which does not contain the direction $\mathbf{e}_1$. The given problem cannot therefore be solved by this method.

It is interesting to see what happens when an attempt is made to solve the same problem (i) using Powell's basic method with a different initial point (Example 4.6); (ii) using Powell's method with the given initial point (Example 4.7, page 137).

Example 4.6

Solve by Powell's basic method:

$$\textit{minimize} \quad f(\mathbf{x}) \equiv x_1^2 - x_1x_2 + 3x_2^2,$$

starting at the point $\mathbf{x}_1 = [2,3]$.

Solution

Iteration 1. $\mathbf{x}_1=[2,3]$, $\mathbf{d}_{11}=\mathbf{e}_1$, $\mathbf{d}_{12}=\mathbf{e}_2$.

Search from $\mathbf{y}_{10}=\mathbf{x}_1$ in the direction $\mathbf{d}_{11}=\mathbf{e}_1$ for a minimum at $\mathbf{y}_{11}=[3/2,3]$. Search from $\mathbf{y}_{11}$ in the direction $\mathbf{d}_{12}=\mathbf{e}_2$ for a minimum at $\mathbf{y}_{12}=[3/2,1/4]$. Search from $\mathbf{y}_{12}$ in the direction $\delta_1=\mathbf{y}_{12}-\mathbf{x}_1=[-1/2,-11/4]$, or equivalently $\delta_1=[-2,-11]$, for a minimum at $\mathbf{x}_2=[512/345,56/345]$.

Iteration 2. $\mathbf{x}_2=[512/345,56/345]$, $\mathbf{d}_{21}=\mathbf{e}_2$, $\mathbf{d}_{22}=\delta_1=[-2,-11]$.

Search from $\mathbf{y}_{20}$ in the direction $\mathbf{d}_{21} = \mathbf{e}_2$ for a minimum at $\mathbf{y}_{21} = [512/345, 256/1035]$. Search from $\mathbf{y}_{21}$ in the direction $\mathbf{d}_{22} = \delta_1$ for a minimum at $\mathbf{y}_{22} = [524288/3(345)^2, 57344/3(345)^2]$. Search from $\mathbf{y}_{22}$ in the direction $\delta_2 = \mathbf{y}_{22} - \mathbf{x}_2 = [-5632/3(345)^2, -616/3(345)^2]$, or equivalently $\delta_2 = [-64, -7]$, for a minimum at $\mathbf{x}_3 = [0,0]$.

This completes iteration 2. Since $f(\mathbf{x})$ is a quadratic function of two variables and the linear searches have been performed exactly, we have reached the optimal point

$$\mathbf{x}^* = \mathbf{x}_3 = [0,0]; \quad f(\mathbf{x}^*) = 0.$$

Powell's method

Powell modified his basic method to overcome the type of difficulty encountered in Example 4.5 by allowing a direction other than $\mathbf{d}_{k1}$ to be discarded after the kth iteration. In this way, the n directions of search can be chosen so as to be always linearly independent; in some cases, the same n directions are used for two successive iterations. Unfortunately, the modification sometimes allows one of the mutually conjugate directions to be discarded, so that more than n iterations are required in order to find the minimum of a positive definite quadratic function. However, the modification is essential if $f(\mathbf{x})$ is a function of more than, say, five variables, and the method has then been found satisfactory for functions of up to fifteen variables.

The kth iteration starts with the current point $\mathbf{x}_k$ and the n directions $\mathbf{d}_{kj}$, and proceeds as follows:

1. Let $\mathbf{y}_{k0} = \mathbf{x}_k$, and for $r = 1, \ldots, n$ search from $\mathbf{y}_{k,r-1}$ in the direction $\mathbf{d}_{kr}$ for a minimum at $\mathbf{y}_{kr}$.
2. Find

$$\begin{aligned} \Delta &= \max_{r=1,\ldots,n} \{f(\mathbf{y}_{k,r-1}) - f(\mathbf{y}_{kr})\} \\ &= f(\mathbf{y}_{k,q-1}) - f(\mathbf{y}_{kq}), \end{aligned}$$

i.e. q is the value of r which maximizes Δ.

3. Define

$$f_1 = f(\mathbf{y}_{k0}), \quad f_2 = f(\mathbf{y}_{kn}),$$

and evaluate

$$f_3 = f(2\mathbf{y}_{kn} - \mathbf{y}_{k0}).$$

4. If either

$$f_3 \geqslant f_1$$

or

$$(f_1 - 2f_2 + f_3)(f_1 - f_2 - \Delta)^2 \geqslant \tfrac{1}{2}\Delta(f_1 - f_3)^2,$$

use the old directions $\mathbf{d}_{kj}$ for the $(k+1)$th iteration and put

$$\mathbf{x}_{k+1} = \mathbf{y}_{kn} = \mathbf{y}_{k+1,0}.$$

Otherwise, use rule 5.

5. Determine $\mathbf{x}_{k+1}$ as in the kth iteration of Powell's basic method (page 129), but instead of the directions (4.56) take for the $(k+1)$th iteration the directions

$$\mathbf{d}_{k1},\ldots,\mathbf{d}_{k,q-1},\mathbf{d}_{k,q+1},\ldots,\mathbf{d}_{kn},\boldsymbol{\delta}_k. \tag{4.58}$$

[If rule 2 gives $q = 1$, then the directions (4.58) are the same as the directions (4.56) of Powell's basic method.]

The convergence of the method is discussed on page 137. Rule 1 is unchanged from Powell's basic method. The theory of Powell's method, resulting in rules 2 to 5 above, is contained largely in the following two theorems.

***Theorem* 4.7**

If the vectors $\mathbf{d}_j$ *are scaled so that*

$$\tfrac{1}{2}\mathbf{d}_j'\mathbf{G}\mathbf{d}_j = 1, \tag{4.59}$$

where $\mathbf{G}$ *is a positive definite symmetric matrix, then the determinant of the matrix whose columns are the* $\mathbf{d}_j$ *takes its maximum value if and only if the directions* $\mathbf{d}_j$ *are mutually conjugate with respect to* $\mathbf{G}$.

Proof

It is possible to choose a set of mutually conjugate directions $\mathbf{y}_j$ satisfying

$$\tfrac{1}{2}\mathbf{y}_k'\mathbf{G}\mathbf{y}_l = \delta_{kl} \quad (k = 1,\ldots,n; \quad l = 1,\ldots,n), \tag{4.60}$$

where δ_{kl} is the Kronecker symbol. Let $\mathbf{D}$ be the $(n \times n)$ matrix whose columns are the vectors $\mathbf{d}_j$ and let $\mathbf{Y}$ be the $(n \times n)$ matrix whose columns are the vectors $\mathbf{y}_j$. Then there exists an $(n \times n)$ matrix $\mathbf{P} = \{p_{jk}\}$ such that

$$\mathbf{D} = \mathbf{YP}, \tag{4.61}$$

since the directions $\mathbf{y}_j$ are mutually conjugate and therefore linearly independent. Hence

$$\det \mathbf{D} = (\det \mathbf{Y})(\det \mathbf{P}).$$

We regard $\mathbf{Y}$ as fixed and we attempt to maximize $\det \mathbf{D}$ by maximizing $\det \mathbf{P}$.

Equation (4.61) may be written in the form

$$\mathbf{d}_j = \sum_{k=1}^{n} p_{kj}\mathbf{y}_k,$$

and hence

$$\tfrac{1}{2}\mathbf{d}_r'\mathbf{G}\mathbf{d}_s = \tfrac{1}{2}\sum_{k=1}^{n}\sum_{l=1}^{n} p_{kr}p_{ls}\mathbf{y}_k'\mathbf{G}\mathbf{y}_l$$

$$= \sum_{k=1}^{n} p_{kr}p_{ks}, \tag{4.62}$$

using equations (4.60). In particular, if $r = s$, we have

$$\tfrac{1}{2}\mathbf{d}_r'\mathbf{G}\mathbf{d}_r = \sum_{k=1}^{n} p_{kr}^{\;2} = 1 \quad (r = 1,\ldots,n), \tag{4.63}$$

where we have used equations (4.59). Hence, by a well-known theorem of Hadamard[77] (see also Exercise 9, Chapter 1), we have

$$\det \mathbf{P} \leqslant 1,$$

the equality holding if and only if $\mathbf{P}$ is an orthogonal matrix. In the present case, $\mathbf{P}$ is orthogonal if and only if

$$\tfrac{1}{2}\mathbf{d}_r'\mathbf{G}\mathbf{d}_s = \delta_{rs}, \tag{4.64}$$

on using equations (4.62) and (4.63) together with the fundamental property of orthogonal matrices that

$$\sum_{k=1}^{n} p_{kr}p_{ks} = 0 \quad (r \neq s).$$

Taking equations (4.59) into account, we see that equations (4.64) hold if and only if the directions $\mathbf{d}_j$ are mutually conjugate, and the theorem is proved.

Powell applies Theorem 4.7 by using the new direction $\boldsymbol{\delta}_k$, derived from the kth iteration, if it can cause an increase in the value of the determinant whose columns are the $\mathbf{d}_{kj}$. The direction to be discarded, if any, is chosen by the condition that the value of this determinant is made as large as possible. Theorem 4.8 shows that these criteria lead to rule 2 above.

Note. In the remainder of this section we shall use the abbreviated notation $\mathbf{y}_j$ for $\mathbf{y}_{kj}$, $\mathbf{d}_j$ for $\mathbf{d}_{kj}$ and $\boldsymbol{\delta}$ for $\boldsymbol{\delta}_k$.

***Theorem* 4.8**

In Powell's method, the direction to be discarded, if any, after the kth iteration is $\mathbf{d}_q$, where q is given by rule 2, *page* 131.

Proof

Suppose the function to be minimized is

$$f(\mathbf{x}) \equiv \tfrac{1}{2}\mathbf{x}'\mathbf{G}\mathbf{x} + \mathbf{b}'\mathbf{x} + c,$$

where $\mathbf{G}$ is positive definite and symmetric. In the kth iteration, the value of λ_r such that

$$f(\mathbf{y}_r) = \tfrac{1}{2}(\mathbf{y}_{r-1} + \lambda_r\mathbf{d}_r)'\mathbf{G}(\mathbf{y}_{r-1} + \lambda_r\mathbf{d}_r) + \mathbf{b}'(\mathbf{y}_{r-1} + \lambda_r\mathbf{d}_r) + c \tag{4.65}$$

is a minimum is given by

$$\partial f(\mathbf{y}_r)/\partial\lambda_r = 0,$$

i.e.

$$\mathbf{d}_r'\mathbf{G}(\mathbf{y}_{r-1} + \lambda_r\mathbf{d}_r) + \mathbf{b}'\mathbf{d}_r = 0.$$

If $\frac{1}{2}\mathbf{d}_r'\mathbf{G}\mathbf{d}_r = 1$, this minimizing value of λ_r is

$$\lambda_r^* = -\tfrac{1}{2}\mathbf{d}_r'\mathbf{G}\mathbf{y}_{r-1} - \tfrac{1}{2}\mathbf{b}'\mathbf{d}_r. \tag{4.66}$$

From equation (4.65), the improvement in the objective function in the current step is

$$f(\mathbf{y}_{r-1}) - f(\mathbf{y}_r) = -[\lambda_r^*\mathbf{d}_r'\mathbf{G}\mathbf{y}_{r-1} + \lambda_r^{*2} + \mathbf{b}'\lambda_r^*\mathbf{d}_r], \tag{4.67}$$

since

$$f(\mathbf{y}_{r-1}) = \tfrac{1}{2}\mathbf{y}_{r-1}'\mathbf{G}\mathbf{y}_{r-1} + \mathbf{b}'\mathbf{y}_{r-1} + c.$$

From equations (4.66) and (4.67), we now obtain

$$f(\mathbf{y}_{r-1}) - f(\mathbf{y}_r) = \lambda_r^{*2}. \tag{4.68}$$

Hence the displacement from $\mathbf{y}_{r-1}$ to $\mathbf{y}_r$ is

$$\lambda_r^*\mathbf{d}_r = [f(\mathbf{y}_{r-1}) - f(\mathbf{y}_r)]^{1/2}\mathbf{d}_r.$$

The new direction derived from the kth iteration is

$$\mathbf{y}_n - \mathbf{y}_0 = \sum_j \lambda_j^*\mathbf{d}_j.$$

If we put

$$\mathbf{y}_n - \mathbf{y}_0 = \mu\delta, \tag{4.69}$$

where

$$\tfrac{1}{2}\delta'\mathbf{G}\delta = 1,$$

then the effect of replacing the direction $\mathbf{d}_r$ by δ is to multiply the determinant of directions by λ_r^*/μ. This follows from the equation

$$\delta = \sum_j (\lambda_j^*/\mu)\mathbf{d}_j$$

and the well-known theorem on determinants concerning the replacement of a column by a linear combination of columns. Hence the direction to be discarded, if any, is that for which λ_r^* is largest, and, because of equation (4.68), this is the direction $\mathbf{d}_q$, where q is given by rule 2, page 131.

It remains to justify rules 3, 4 and 5, pp. 131-2. The proof of Theorem 4.8 shows that δ should replace $\mathbf{d}_q$ if and only if $\lambda_q^* > \mu$. To calculate μ, the function values f_1, f_2 and f_3 of rule 3 are used. Because these function values occur at equally spaced points along the direction δ, the predicted stationary value of $f(\mathbf{x})$ along this direction is

$$f_s = f_2 - \frac{(f_1 - f_3)^2}{8(f_1 - 2f_2 + f_3)}, \tag{4.70}$$

from equation (3.21). Also, this value is a minimum if

$$M \equiv f_1 - 2f_2 + f_3 > 0.$$

If $M < 0$, then a new direction should certainly be defined, since equation (4.70) shows that the current n directions give $f_s > f_2$, which is clearly unsatisfactory.

If $M > 0$, then f_s is a minimum and it is necessary to consider separately the cases $f_3 \geqslant f_1$ and $f_3 < f_1$.

(a) $f_3 \geqslant f_1$.

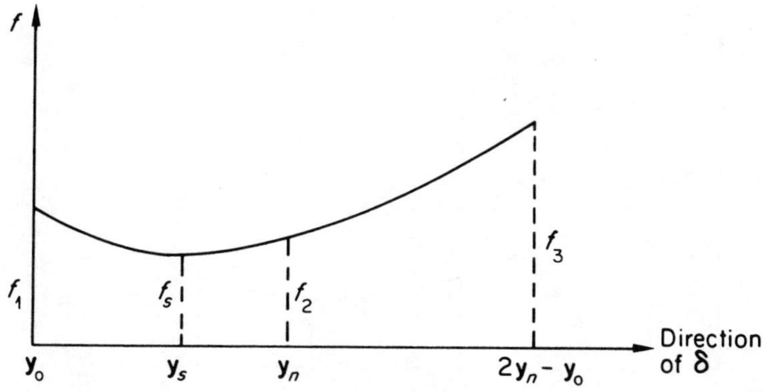

Figure 4.3 Powell's method, $f_3 \geqslant f_1$

Figure 4.3 shows diagrammatically the values of the quadratic function $f(\mathbf{x})$ along the direction δ from the point $\mathbf{y}_0$. Let

$$\mathbf{y}_s - \mathbf{y}_0 = \mu_1 \delta, \quad \mathbf{y}_n - \mathbf{y}_s = \mu_2 \delta.$$

Then

$$\mathbf{y}_n - \mathbf{y}_0 = (\mu_1 + \mu_2)\delta = \mu\delta,$$

using equation (4.69). Hence

$$\mu = \mu_1 + \mu_2;$$

but, by reasoning similar to that which led to equation (4.68), we find

$$\mu_1 = (f_1 - f_s)^{1/2}, \quad \mu_2 = (f_2 - f_s)^{1/2},$$

and therefore

$$\mu = (f_1 - f_s)^{1/2} + (f_2 - f_s)^{1/2}. \tag{4.71}$$

From rule 2, page 131, and equation (4.68), we have

$$\Delta^{1/2} = \lambda_q{}^* = \max_{r=1,\ldots,n} \{[f(\mathbf{y}_{r-1}) - f(\mathbf{y}_r)]^{1/2}\}. \tag{4.72}$$

Since

$$f_1 = f(\mathbf{y}_0) > f(\mathbf{y}_1) > \ldots. > f(\mathbf{y}_s) = f_s$$

and

$$f_2 = f(\mathbf{y}_n) > f(\mathbf{y}_{n-1}) > .... > f(\mathbf{y}_s) = f_s,$$

it is clear from equations (4.71) and (4.72) that $\mu > \lambda_q{}^*$. This means that the old directions should be used again, which verifies the first part of rule 4, page 131.

(b) $f_3 < f_1$.

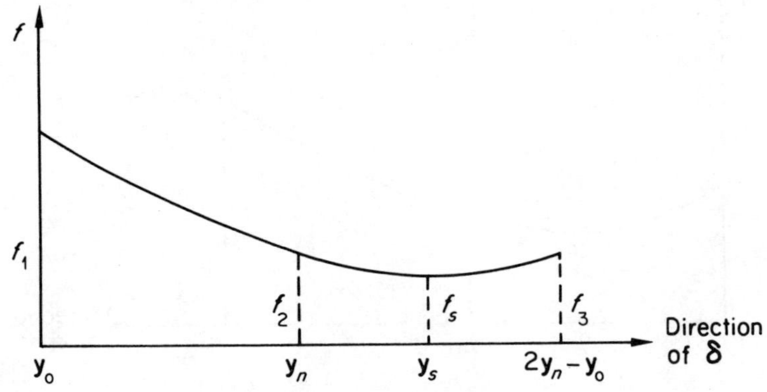

Figure 4.4 Powell's method, $f_3 < f_1$

In this case, $\mathbf{y}_n$ lies between $\mathbf{y}_0$ and $\mathbf{y}_s$, as shown in Figure 4.4. The reasoning is similar to that of case (a). Let

$$\mathbf{y}_s - \mathbf{y}_0 = \mu_1 \delta, \quad \mathbf{y}_s - \mathbf{y}_n = \mu_2 \delta.$$

Then

$$\mathbf{y}_n - \mathbf{y}_0 = (\mu_1 - \mu_2)\delta = \mu\delta.$$

Hence

$$\mu = \mu_1 - \mu_2;$$

but

$$\mu_1 = (f_1 - f_s)^{1/2}, \quad \mu_2 = (f_2 - f_s)^{1/2},$$

and therefore

$$\mu = (f_1 - f_s)^{1/2} - (f_2 - f_s)^{1/2}.$$

Since $\lambda_q{}^* = \Delta^{1/2}$, where Δ is defined in rule 2, page 131, the condition $\lambda_q{}^* > \mu$ that new directions should be used becomes

$$\Delta^{1/2} > (f_1 - f_s)^{1/2} - (f_2 - f_s)^{1/2}. \tag{4.73}$$

The old directions are used again if the inequality (4.73) is *not* satisfied, i.e. if

$$\Delta^{1/2} \leqslant (f_1 - f_s)^{1/2} - (f_2 - f_s)^{1/2}. \tag{4.74}$$

Eliminating the surds and using equation (4.70) we find, after some manipulation, that the inequality (4.74) becomes

$$\frac{\frac{1}{2}\Delta(f_1 - f_3)^2}{f_1 - 2f_2 + f_3} \leqslant (f_1 - f_2 - \Delta)^2,$$

which is equivalent to

$$(f_1 - 2f_2 + f_3)(f_1 - f_2 - \Delta)^2 \geqslant \tfrac{1}{2}\Delta(f_1 - f_3)^2, \tag{4.75}$$

since

$$f_1 - 2f_2 + f_3 > 0. \tag{4.76}$$

The inequality (4.75) verifies the second part of rule 4, page 131. Note that if either of the conditions of rule 4 is satisfied, then the necessary condition (4.76) for f_s to be a minimum is automatically satisfied. This is obvious for the second of these conditions, namely (4.75); also, since $f_2 < f_1$, the first condition $f_3 \geqslant f_1$ implies (4.76). The justification of rules 3, 4 and 5 is now complete.

The rate of convergence of the method is usually acceptable, since the value of the determinant whose columns represent the search directions cannot decrease. An obvious criterion for terminating the calculations is that an iteration should cause each variable to change by less than some prescribed quantity ε, i.e.

$$|x_{k+1,j} - x_{kj}| < \varepsilon.$$

Although this criterion is usually successful, there are occasions when it terminates the calculations prematurely; this can happen when a direction exists along which $f(\mathbf{x})$ varies very slowly. The original paper by Powell[62] gives details of a very safe convergence criterion which was used when the method was coded as a general sub-routine.

Example 4.7

Solve by Powell's method:

$$\textit{minimize} \quad f(\mathbf{x}) \equiv x_1{}^2 - x_1 x_2 + 3x_2{}^2,$$

starting at the point $\mathbf{x}_1 = [1,2]$.

Solution

Iteration 1. $\quad \mathbf{x}_1 = [1,2], \ \mathbf{d}_{11} = \mathbf{e}_1, \ \mathbf{d}_{12} = \mathbf{e}_2.$

1. Search from $\mathbf{y}_{10} = \mathbf{x}_1$ in the direction $\mathbf{d}_{11}$ for a minimum at $\mathbf{y}_{11} = [1,2]$. Search from $\mathbf{y}_{11}$ in the direction $\mathbf{d}_{12}$ for a minimum at $\mathbf{y}_{12} = [1,1/6]$.
2. $\Delta = \max\{f(\mathbf{y}_{10}) - f(\mathbf{y}_{11}), f(\mathbf{y}_{11}) - f(\mathbf{y}_{12})\}$
   $= \max\{11 - 11, 11 - \frac{11}{12}\} = 121/12$, i.e. $q = 2$.
3. $f_1 = f(\mathbf{y}_{10}) = 11$, $f_2 = f(\mathbf{y}_{12}) = 11/12$, $f_3 = f(2\mathbf{y}_{12} - \mathbf{y}_{10}) = 11$.
4. $f_3 = f_1$; we therefore use the old directions $\mathbf{d}_{11}, \mathbf{d}_{12}$ in the second iteration.

Iteration 2. $\mathbf{x}_2 = \mathbf{y}_{12} = [1, 1/6]$, $\mathbf{d}_{21} = \mathbf{e}_1$, $\mathbf{d}_{22} = \mathbf{e}_2$.

1. Search from $\mathbf{y}_{20} = \mathbf{x}_2$ in the direction $\mathbf{d}_{21}$ for a minimum at $\mathbf{y}_{21} = [1/12, 1/6]$. Search from $\mathbf{y}_{21}$ in the direction $\mathbf{d}_{22}$ for a minimum at $\mathbf{y}_{22} = [1/12, 1/72]$.
2. $\Delta = \max\{f(\mathbf{y}_{20}) - f(\mathbf{y}_{21}), f(\mathbf{y}_{21}) - f(\mathbf{y}_{22})\}$

$$= \max\left\{\frac{11}{12} - \frac{11}{144}, \frac{11}{144} - \frac{33}{5184}\right\}$$

$$= \max\left\{\frac{121}{144}, \frac{121}{1728}\right\} = \frac{121}{144}, \quad \text{i.e. } q = 1.$$

3. $f_1 = f(\mathbf{y}_{20}) = 11/12$, $f_2 = f(\mathbf{y}_{22}) = 33/5184$, $f_3 = f(2\mathbf{y}_{22} - \mathbf{y}_{20}) = 275/432$.
4. $f_3 < f_1$,

$$(f_1 - 2f_2 + f_3)(f_1 - f_2 - \Delta)^2 < \tfrac{1}{2}\Delta(f_1 - f_3)^2.$$

We therefore proceed to rule 5.

5. Search from $\mathbf{y}_{22}$ in the direction

$$\delta_2 = \mathbf{y}_{22} - \mathbf{x}_2 = \left[-\frac{11}{12}, -\frac{11}{72}\right]$$

for a minimum at $\mathbf{x}_3 = [0,0]$. The directions for the third iteration are $\mathbf{d}_{22}$, δ_2, but it is obvious that $f(\mathbf{x})$ has a minimum in *any* direction at the point $[0,0]$. Hence

$$\mathbf{x}^* = [0,0], \quad f(\mathbf{x}^*) = 0.$$

4.10 CHOICE OF METHOD

In this section, we consider briefly the question of the choice of method, beginning with the case where it is possible to evaluate the derivatives $\partial f/\partial x_j$ without undue difficulty or expense. For the minimization of a general function of any number of variables, there is evidence to show that the complementary DFP method is preferable to the DFP method—see the article by Fletcher (Chapter 8, 'A Survey of Algorithms for Unconstrained Optimization').[55] Given sufficient computer storage, one of these two methods should be tried first; with less computer storage available the Fletcher–Reeves method should be tried.

The method of steepest descent is not recommended for general use because of its poor convergence properties. The Newton–Raphson method has much better convergence properties; it works particularly well if a close initial estimate of the optimal point can be found. However, it may fail to converge from a poor initial estimate of the optimal point. Also, the evaluation of the elements of the Hessian matrix, and the inversion of this matrix, may pose formidable computational problems.

In the cases where it is difficult or impossible to evaluate the derivatives $\partial f/\partial x_j$, the DFP or complementary DFP method can still be used, with the derivatives approximated by differences. Alternatively, Powell's method may be used, although this method tends to lose its efficiency if the number of variables exceeds fifteen or so, since there is a tendency for new directions of search to be chosen less often as the number of variables increases. Smith's method is among the simplest of this type but has been superseded by the above algorithms. The possibility of using a direct search method should not be overlooked.

A very useful review paper by Fletcher[32] uses seven different test functions to compare three minimization algorithms that do not require the evaluation of derivatives, namely, the methods of Davies, Swann and Campey,[74] Powell[62] and Smith.[72]

In 1970, Huang[47] introduced a large three-parameter family of rank 2 algorithms with quadratic termination, which included as special cases the DFP, complementary DFP and Fletcher–Reeves algorithms. The relative merits of these last three algorithms are put into perspective by a theorem proved by Huang for quadratic objective functions. He showed that, given the same initial point and the same initial search direction, together with exact linear searches, the same sequence of current points and search directions is generated by every member of his family of algorithms.

Following some numerical experiments by Huang and Levy,[48] the corresponding results for non-quadratic objective functions were obtained by Dixon.[26] He showed that, under the above conditions, the sequence of current points and search directions generated by members of Huang's family depends on one parameter only. Furthermore, the DFP, complementary DFP and Fletcher–Reeves algorithms all have the same value of this parameter and so produce identical sequences of current points and search directions.

These results suggest that the Fletcher–Reeves algorithm—the simplest algorithm of Huang's family—should be used in preference to the more elaborate DFP and similar algorithms. However, the Fletcher–Reeves algorithm may require high-precision arithmetic and high-accuracy linear searches, and the latter in particular can be very time-consuming. The complementary DFP formula without linear searches and Powell's 1972 method[65] which does not require linear searches should therefore be considered as alternatives.

If $f(\mathbf{x})$ is a sum of squares of functions, as is often the case in data-fitting problems or in the solution of systems of nonlinear equations, then special methods are available, e.g. the methods of Levenberg,[53] Marquardt[54] and Powell.[63] These methods can be expected to converge faster than the general purpose methods considered in this chapter. They are derived from the 'generalized least squares method' which is described at the beginning of Powell's paper.[63] This is an extension of the original 'method of least squares' of Legendre (1805) and Gauss (1809). Levenberg's and Marquardt's methods are similar, though they were discovered independently. They use the first derivatives of the objective function, while Powell's method requires function evaluations only.

A discussion of some practical points in connection with the implementation of Levenberg's and similar methods is given by Beale.[4] Bard[2] compares the efficiencies of the methods that require the evaluation of derivatives for the maximization of a sum of squares. Finally, Powell gives a comprehensive review of least squares algorithms in Chapter 3 ('Problems Related to Unconstrained Optimization').[55]

SUMMARY

At the point $\mathbf{x} = \mathbf{x}_k$, the function $f(\mathbf{x})$ increases most rapidly in the gradient direction $\nabla f(\mathbf{x}_k)$. Gradient methods for minimizing $f(\mathbf{x})$ often use a sequence of linear searches along successive directions. In the method of steepest descent, the search from a current point is carried out along the negative gradient direction at that point. In practice, it is found that better results are obtained by searching along mutually conjugate directions. The theory of gradient methods is highly developed in the case where the objective function $f(\mathbf{x})$ is quadratic in its arguments, but is less highly developed for more general functions $f(\mathbf{x})$. However, the algorithms derived from 'quadratic' theory may be applied iteratively to non-quadratic objective functions. Some of these algorithms, e.g. the DFP method, the complementary DFP method and the Fletcher–Reeves method, are among the most efficient general purpose optimization techniques available at the present time. Not all gradient methods require analytic expressions for the derivatives. The idea of conjugate directions, on which the most successful methods are based, may also be used when the derivatives cannot be evaluated directly. Powell's method is specifically designed to generate conjugate directions of search without the need to evaluate derivatives; also, quasi-Newton methods with the derivatives replaced by differences have been used successfully. Special methods are available for the minimization of a sum of squares.

EXERCISES

1. Solve by the method of steepest descent (a) with fixed step lengths, (b) with linear searches:

$$\text{minimize} \quad f(\mathbf{x}) \equiv 2x_1{}^4 + x_2{}^2 - 4x_1x_2 + 5x_2,$$

taking the origin as the initial point in each case.

2. Solve the problem of Exercise 1 by the Newton–Raphson method, using (a) formula (4.5), (b) formula (4.6).

3. Solve by the Newton–Raphson method:

$$\text{minimize} \quad f(\mathbf{x}) \equiv [1 + (x_1 + x_2 - 5)^2]\,[1 + (3x_1 - 2x_2)^2],$$

taking the initial point to be (a) [10,10], (b) [2,2].

4. Prove that if the matrix $\mathbf{G}$ is positive definite then so is $\mathbf{G}^{-1}$. [This result is used on page 109.]

5. Solve by (a) the DFP method, (b) the complementary DFP method with

linear searches, (c) the complementary DFP method without linear searches, (d) the Fletcher–Reeves method:

$$\text{minimize} \quad f(\mathbf{x}) \equiv 3x_1{}^2 - 2x_1x_2 + x_2{}^2 + x_1,$$

taking [1,1] as the initial point.

6. Use the DFP method to solve the simultaneous equations

$$\begin{aligned} 10x_1 + \;\, 9x_2 + \;\, 7x_3 &= 54, \\ 3x_1 + \;\, 6x_2 + \;\, 4x_3 &= 28, \\ 40x_1 + 52x_2 + 38x_3 &= 275, \end{aligned}$$

with an error of less than 10^{-5} in each component of $\mathbf{x}^*$.

7. Prove that every positive definite symmetric matrix has a square root.

8. In the DFP method, prove that $R(\mathbf{H}_{k+1})$ is a subspace of $R(\mathbf{H}_k)$, where $R(\mathbf{H}_k)$ is the range of $\mathbf{H}_k$. Deduce that if $\mathbf{H}_k$ is singular then the set of all subsequent search directions does not span E^n—hence the optimal point may not be found. [The definition of the range $R(\mathbf{A})$ of a matrix $\mathbf{A}$ is given in Exercise 23, Chapter 1.]

9. Write computer programs for (a) the DFP method, (b) the Fletcher–Reeves method, (c) Powell's method, and use them to minimize

$$f(\mathbf{x}) \equiv [1 + (x_1 + 2x_2 + x_3 - 3)^2]\,[1 + (4x_1 + x_2 + 2x_3 - 5)^2] \times [1 + (2x_1 + 3x_2 - x_3 - 2)^2],$$

taking the origin as the initial point.

10. Two diameters of an ellipse are said to be *conjugate* if each is parallel to the tangents at the extremities of the other. Let

$$y = \mu x, \quad y = \mu' x$$

be conjugate diameters of the ellipse

$$ax^2 + 2hxy + by^2 = 1.$$

Prove that

$$a + h(\mu + \mu') + b\mu\mu' = 0,$$

and show that this condition can be written in the form

$$\mathbf{p}'\mathbf{G}\mathbf{q} = 0,$$

where

$$\mathbf{G} = \begin{pmatrix} a & h \\ h & b \end{pmatrix}, \; \mathbf{p} = [\cos\alpha, \sin\alpha], \;\; \mathbf{q} = [\cos\alpha', \sin\alpha'], \;\; \mu = \tan\alpha, \; \mu' = \tan\alpha'.$$

11. Prove or disprove that if $\mathbf{a}$ is conjugate to $\mathbf{b}$, and $\mathbf{b}$ is conjugate to $\mathbf{c}$, then $\mathbf{a}$ is conjugate to $\mathbf{c}$.

12. Given the quadratic function

$$f(\mathbf{x}) \equiv \tfrac{1}{2}\mathbf{x}'\mathbf{G}\mathbf{x} + \mathbf{b}'\mathbf{x} + c,$$

where **G** is symmetric, prove that

$$f_1 - f_2 = \tfrac{1}{2}\mathbf{u}'\mathbf{v},$$
$$\mathbf{g}_1^{\,2} - \mathbf{g}_2^{\,2} = \mathbf{u}'\mathbf{G}\mathbf{v},$$

where

$$f_1 = f(\mathbf{x}_1),\ \mathbf{g}_1 = \nabla f(\mathbf{x}_1),\ \text{etc.},$$
$$\mathbf{u} = \mathbf{x}_1 - \mathbf{x}_2,$$
$$\mathbf{v} = \mathbf{G}(\mathbf{x}_1 + \mathbf{x}_2) + 2\mathbf{b}.$$

13. Solve by the Fletcher–Reeves method:

$$\text{minimize} \quad f(\mathbf{x}) \equiv 30(x_1^{\,2} - 2000x_2^{\,3})^2 + 7500x_2^{\,4} - x_1,$$

with an error of less than 10^{-4} in $f(\mathbf{x}^*)$.

14. Solve by (a) Smith's method, (b) Powell's basic method:

$$\text{minimize} \quad f(\mathbf{x}) \equiv 3x_1^{\,2} - 4x_1x_2 + 2x_2^{\,2} + 4x_1 + 4,$$

taking the origin as the initial point.

15. Solve by Powell's method:

$$\text{minimize} \quad f(\mathbf{x}) \equiv (x_1 + x_2 + 4)^2 + (x_1 - x_2 + 2)^2,$$

taking the origin as the initial point, with initial directions (a) [1,0], [0,1], (b) [1,0], [1,1].

16. Solve by any method of this chapter:

$$\text{minimize} \quad f(\mathbf{x}) \equiv 3x_1^{\,2} + x_1x_2 + 2x_2^{\,2} - 2x_1 - x_2 + 1.$$

17. The ratio σ of the density of air at height h in the troposphere to its density at sea-level is given theoretically by

$$\sigma = \left(1 - \frac{\alpha h}{T_0}\right)^{(1/\alpha R) - 1},$$

where T_0 is the absolute sea-level temperature (°C), α is the temperature gradient (°C/ft) and R is the gas constant (ft/°C). Use the following data to find optimal values of the parameters T_0 and α, given that $R = 96{\cdot}05$ ft/°C.

h (ft)	2,000	5,000	10,000	20,000	30,000	35,000
σ	0·9427	0·8616	0·7384	0·5326	0·3739	0·3096

CHAPTER 5

Constrained Optimization

5.1 INTRODUCTION

We now consider how the techniques for unconstrained optimization, described in the previous two chapters, can be extended to allow for the presence of constraints. We shall see that most of the methods for dealing with constraints can be used in conjunction with *any* method for unconstrained optimization. A general aim when dealing with a constrained optimization problem is to reduce it to an unconstrained problem or to a sequence of such problems. We shall not discuss direct search methods specifically in this chapter, since it is relatively easy to adapt them to include constraints—see, for example, Exercise 4, Chapter 3.

It is sometimes possible to eliminate one or more of the constraints by a change of variables. For example, the constraint $x_1 \leqslant c$ is equivalent to

$$x_1 = c - X_1{}^2$$

or to

$$x_1 = c - e^{X_1},$$

with X_1 unconstrained. Similarly, $a \leqslant x_1 \leqslant b$ is equivalent to

$$x_1 = a + (b - a)\sin^2 X_1,$$

with X_1 unconstrained. Hence the above inequality constraints can be eliminated by substituting for x_1 wherever it appears. With a little ingenuity, this method can be used for problems with two or more variables; in practice, however, it is useful only in a few fairly simple cases. Box[9] gives some examples.

When constraints are present, a preliminary calculation is often necessary in order to find an initial feasible point. The following method may be used. Starting at some non-feasible point, write the constraints in the form

$$\begin{aligned} g_i(\mathbf{x}) &\geqslant 0 \quad (i = 1, \ldots, v) \\ &= 0 \quad (i = v+1, \ldots, m) \end{aligned}$$

and minimize the function

$$Z = -\sum_{i=1}^{v}{}' g_i(\mathbf{x}) + \sum_{i=v+1}^{m} [g_i(\mathbf{x})]^2,$$

where Σ' indicates that the summation is taken over only those constraints that are violated at the current (non-feasible) point. The minimization of Z is otherwise unconstrained. When the value of Z has been reduced to zero, a feasible point has been found.

An obvious way of dealing with equality constraints is to use them to eliminate one or more variables from the given problem. This procedure, however, may lead to unnecessarily complicated calculations and must therefore be used with caution.

In the following sections, we shall assume that all the optimization problems have been formulated as minimizing problems.

5.2 HEMSTITCHING

This simple method with a picturesque name was devised by Roberts and Lyvers[68] in 1961. It may be used for problems with any number of variables and constraints, though for ease of description we shall consider first a problem with only two variables and one constraint.

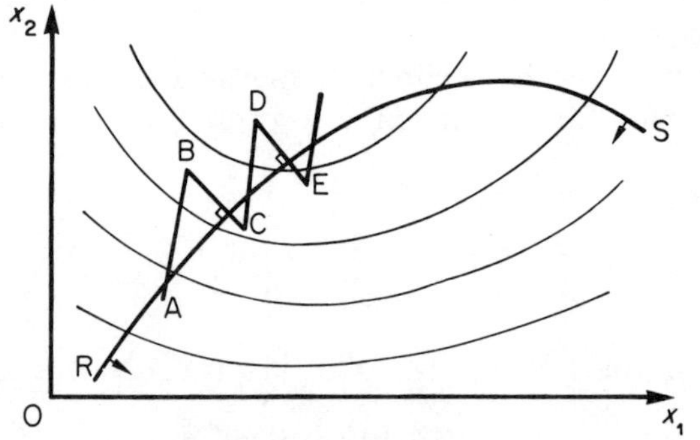

Figure 5.1 Hemstitching

The thin lines in Figure 5.1 are the level curves for the objective function, and RS is the constraint boundary, with the arrows pointing into the feasible region. Let A be the current point and suppose that the next step of an unconstrained minimization algorithm takes the current point to B, thereby violating the constraint. The hemstitching procedure is to take a step from B in a direction orthogonal to the constraint boundary, reaching a feasible point C. The unconstrained minimization is then continued; any further violations of the constraint are corrected in the same way, giving (typically) the hemstitching path ABCDE..... The method requires the calculation of the return direction, i.e. the direction of the normal to the constraint from the non-feasible point. A test is also required to ensure that the new current point is feasible.

When two or more constraints are violated simultaneously, the return direction is given by

$$\lambda_1\mathbf{n}_1 + \lambda_2\mathbf{n}_2 + \ldots.,$$

where $\mathbf{n}_1,\mathbf{n}_2,\ldots.$ are unit normals, or approximate unit normals, to the violated constraints, and $\lambda_1,\lambda_2,\ldots.$ are positive weighting factors which are proportional to the amounts by which the constraints are violated.

A modification to the basic hemstitching method was introduced by Davies.[10,25] In Figure 5.2, where the level curves and the constraint boundary

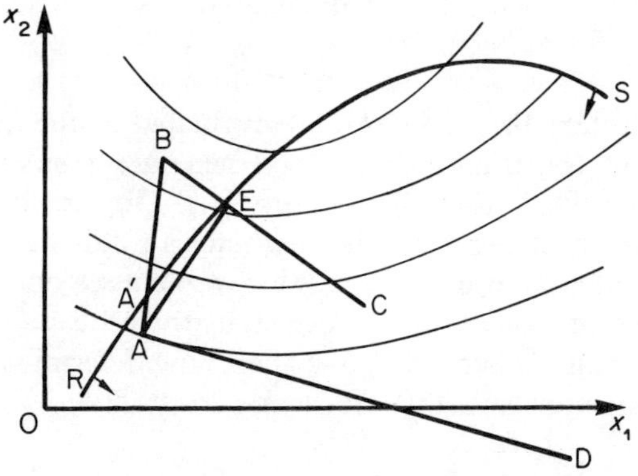

Figure 5.2 Modified hemstitching

are the same as in Figure 5.1, suppose that the current point A is on or very close to the constraint boundary RS. It is again assumed that a step from A in the direction determined by an unconstrained minimization algorithm takes the current point to the non-feasible position B. The line AD is tangential to the level curve through A. [In n dimensions, AD represents the tangent hyperplane at A to the level hypersurface $f(\mathbf{x}) = f_A$.]

If $f(\mathbf{x})$ is a convex function, and if the return point C lies on or beyond AD, then $f_C \geqslant f_A$. For a minimizing problem, the return direction from B is therefore constructed as follows—whether $f(\mathbf{x})$ is convex or not. Let AB intersect the constraint boundary at A'; draw BC parallel to the direction of the normal to the constraint boundary at A' and let E be the point on BC which is just feasible. The point E is found by interpolation between B (non-feasible) and C (feasible). If $f_E < f_A$, continue with the unconstrained minimization algorithm from E. If $f_E \geqslant f_A$, search along AE for the minimizing point, which is certainly feasible if the feasible region is locally convex. Continue with the unconstrained algorithm from this point or from the nearest feasible point. Reduce the step length AB if the return direction fails to intersect the constraint boundary.

For a non-convex objective function, the method may break down when the function value at the new feasible point is greater than f_A. In this case, it may be possible to continue hemstitching with a reduced step length; otherwise a different method must be used, e.g. the gradient projection method of the next section.

5.3 THE GRADIENT PROJECTION METHOD

This method is based on a general nonlinear programming algorithm due to Rosen.[69,70] The current point is assumed to lie on one or more of the constraint boundaries and the basic idea of the method is to search in a direction in which $f(\mathbf{x})$ decreases but which is 'tangential' to the boundaries of the active constraints. Specifically, let the constraints be

$$g_i(\mathbf{x}) \geqslant 0.$$

The gradient vectors $\nabla g_1, \ldots, \nabla g_l$ ($l \leqslant m$), evaluated at the current point, of the active constraint functions $g_1(\mathbf{x}), \ldots, g_l(\mathbf{x})$ generate a subspace S of E^n. It is assumed for simplicity that the vectors $\nabla g_1, \ldots, \nabla g_l$ are linearly independent. If they are linearly dependent, the method can still be used by selecting a linearly independent subset of them; this case causes no difficulty in Rosen's algorithm. The direction of search is given by the projection of $-\nabla f(\mathbf{x})$ on $O(S)$, the orthogonal complement of S. We shall now determine this direction.

If s is any direction in S then, for some constants μ_k,

$$\mathbf{s} = \sum_{k=1}^{l} \mu_k \nabla g_k = \mathbf{N}\boldsymbol{\mu}, \text{ say,} \tag{5.1}$$

where $\boldsymbol{\mu} = [\mu_1, \ldots, \mu_l]$. It is now assumed, without loss of generality, that the ∇g_k are unit vectors (the values of the μ_k may be suitably adjusted). Then the matrix $\mathbf{N}$, of order $(n \times l)$ and rank l, is the matrix of unit column vectors $\nabla g_k = [g_{1k}, \ldots, g_{nk}]$, i.e.

$$\mathbf{N} = \begin{pmatrix} g_{11} & \cdots & g_{1l} \\ & & \\ g_{n1} & \cdots & g_{nl} \end{pmatrix},$$

where

$$\sum_j g_{jk}{}^2 = 1 \quad (k = 1, \ldots, l).$$

Let $\mathbf{g} = \nabla f(\mathbf{x})$. Then we can write

$$\mathbf{g} = \mathbf{r} + \mathbf{s}, \tag{5.2}$$

where $\mathbf{r} \in O(S)$, $\mathbf{s} \in S$, and $-\mathbf{r}$ is the required direction of search. From equations (5.1) and (5.2),

$$\mathbf{g} = \mathbf{r} + \mathbf{N}\boldsymbol{\mu}. \tag{5.3}$$

Since $\mathbf{r} \in O(S)$, we have

$$(\nabla g_k)'\mathbf{r} = 0 \quad (k = 1, \ldots, l),$$

which can be written in the form

$$\mathbf{N}'\mathbf{r} = 0. \tag{5.4}$$

Hence, from equations (5.3) and (5.4),

$$\mathbf{N}'\mathbf{g} = \mathbf{N}'\mathbf{N}\boldsymbol{\mu},$$

and the matrix $\mathbf{N}'\mathbf{N}$ is non-singular because the columns of $\mathbf{N}$ are linearly independent.[27] Thus

$$\boldsymbol{\mu} = (\mathbf{N}'\mathbf{N})^{-1}\mathbf{N}'\mathbf{g}. \tag{5.5}$$

Now, from equations (5.1) and (5.5),

$$\mathbf{s} = \mathbf{N}(\mathbf{N}'\mathbf{N})^{-1}\mathbf{N}'\mathbf{g} \tag{5.6}$$

and, combining equations (5.2) and (5.6), we find

$$\begin{aligned} \mathbf{r} &= \mathbf{g} - \mathbf{N}(\mathbf{N}'\mathbf{N})^{-1}\mathbf{N}'\mathbf{g} \\ &= \mathbf{P}\mathbf{g}, \end{aligned} \tag{5.7}$$

where

$$\mathbf{P} = \mathbf{I} - \mathbf{N}(\mathbf{N}'\mathbf{N})^{-1}\mathbf{N}'.$$

Thus the direction of search is given by

$$-\mathbf{r} = -\mathbf{P}\mathbf{g}.$$

The matrix $\mathbf{P}$ is called a *projection matrix*: it projects the direction on which it operates into the orthogonal complement of the directions represented by the columns of $\mathbf{N}$. In practice, $\mathbf{P}$ is used whenever the current point is on or very close to one or more of the constraint boundaries.

Since

$$-\mathbf{r}'\nabla f(\mathbf{x}) = -\mathbf{r}'\mathbf{g} = -\mathbf{r}'(\mathbf{r} + \mathbf{s}) = -\mathbf{r}'\mathbf{r} < 0,$$

the objective function $f(\mathbf{x})$ decreases in the direction of search $-\mathbf{r}$. The current point is moved in the direction $-\mathbf{r}$ until either $f(\mathbf{x})$ is a minimum in this direction or a new constraint boundary is reached. It sometimes happens that the current point becomes non-feasible during the course of the calculations—this case is considered below.

An important special case arises when every constraint is linear, for the gradient projection method then moves the current point along the constraint boundary in the direction in which $f(\mathbf{x})$ decreases. In this case, the current constraint boundary may be regarded as the intersection of a number of hyperplanes. In each iteration, this intersection consists of a set of hyperplanes which only differs from the previous set by one: either a hyperplane is dropped from the set or one is added. Rosen's algorithm takes advantage of this by providing two recursion relations to compute $(\mathbf{N}'\mathbf{N})^{-1}$, one for discarding and the other for adding a hyperplane. Rosen's original paper[69] should be consulted for details of the procedure.

When the current point becomes non-feasible, it is necessary to compute a return direction. At a non-feasible point, suppose that the absolute values of the constraint functions for the violated constraints are $w_1, \ldots, w_l$. The return direction $\mathbf{s} \in S$ is determined as follows. Let

$$w_k = (\nabla g_k)'\mathbf{s} = |\mathbf{s}| \cos \theta_k \quad (k = 1, \ldots, l), \tag{5.8}$$

where the unit vectors ∇g_k are as defined earlier and θ_k is the angle between ∇g_k and $\mathbf{s}$. It is clear from equations (5.8) that if a particular w_k, say w_r, increases relative to the remaining w_k, then the direction of $\mathbf{s}$ approaches that of ∇g_r.

Equations (5.8) can be written

$$\mathbf{w} = \mathbf{N}'\mathbf{s},$$

where $\mathbf{w} = [w_1, \ldots, w_l]$. Using equation (5.1), we obtain

$$\mathbf{w} = \mathbf{N}'\mathbf{N}\boldsymbol{\mu},$$

and hence

$$\boldsymbol{\mu} = (\mathbf{N}'\mathbf{N})^{-1}\mathbf{w}. \tag{5.9}$$

Finally, from equations (5.1) and (5.9), the return direction is

$$\mathbf{s} = \mathbf{N}(\mathbf{N}'\mathbf{N})^{-1}\mathbf{w}.$$

A computational advantage of the method is that the return direction $\mathbf{s}$ is easily found once the matrix $(\mathbf{N}'\mathbf{N})^{-1}$ is known, and the latter has already been evaluated in determining the projection matrix $\mathbf{P}$.

We have considered gradient projection as a method for dealing with constraints, to be used in conjunction with any unconstrained minimization algorithm. This account does less than justice to the gradient projection method as described by Rosen,[69,70] for in its complete form it is a general purpose method for solving both linear and nonlinear programming problems.

5.4 PENALTY FUNCTIONS

A penalty function is a function $P(\mathbf{x})$ which is chosen so that the optimal solution of the constrained problem:

$$\text{minimize} \quad z = f(\mathbf{x}),$$

subject to one or more constraints on $\mathbf{x}$, is close to the optimal solution of the unconstrained problem:

$$\text{minimize} \quad Z = f(\mathbf{x}) + P(\mathbf{x}).$$

The following examples illustrate some commonly used penalty functions. Note that in each case the 'penalty' for violating a constraint is a high value of Z, the modified objective function.

(a) To confine x to the interval $[-X, X]$, we may use

$$P(x) = k\left(\frac{x}{X}\right)^{2M}, \qquad k > 0,\ M \text{ a positive integer,} \tag{5.10}$$

or, alternatively,

$$P(x) = -k\ln\left(\cos\frac{\pi x}{2X}\right), \quad k > 0. \tag{5.11}$$

More generally, if the constraint is $X_1 \leqslant x \leqslant X_2$, the penalty functions (5.10) and (5.11) are replaced respectively by

$$P(x) = k\left[\frac{2x - (X_1 + X_2)}{X_2 - X_1}\right]^{2M}, \quad k > 0,\ M \text{ a positive integer,}$$

and

$$P(x) = -k\ln\left(\cos\left[\frac{\pi\{2x - (X_1 + X_2)\}}{X_2 - X_1}\right]\right), \quad k > 0.$$

(b) The inequality constraints

$$g_i(\mathbf{x}) \geqslant 0$$

may be replaced by

$$P(\mathbf{x}) = \sum_i k_i[g_i(\mathbf{x})]^2, \qquad k_i > 0 \text{ for all } i, \tag{5.12}$$

or, more generally, by

$$P(\mathbf{x}) = \sum_i k_i[g_i(\mathbf{x})]^{2M}, \qquad k_i > 0 \text{ for all } i,\ M \text{ a positive integer,} \tag{5.13}$$

where the summation in each case is taken over the violated constraints.

(c) The equality constraints

$$g_i(\mathbf{x}) = 0$$

may be replaced by the penalty functions (5.12) or (5.13), where the summations now extend over all the constraints. Alternatively, we may use

$$P(\mathbf{x}) = \sum_i k_i|g_i(\mathbf{x})|, \qquad k_i > 0 \text{ for all } i.$$

The method of penalty functions is simple and effective, provided that suitable values of the parameters can be chosen—some numerical trial and error is often necessary.

Geometrically, a penalty function replaces a constraint by a steep-sided ridge (minimizing problem) or valley (maximizing problem), and may therefore cause difficulties if gradient methods are used. It should also be noted that function evaluations may be needed at non-feasible points. The sequential unconstrained minimization technique (Section 5.5) is a highly developed form of the penalty function method in which function evaluations are needed only at feasible points.

The following analysis, given by Kelley,[49] indicates how the parameters k_i should be adjusted as the optimal point is approached. It may be applied to penalty functions which replace both equality and inequality constraints, provided that the index i is taken over all the equality constraints as well as the violated inequality constraints. Suppose that the unconstrained problem is:

$$\text{minimize} \quad Z(\mathbf{x}) \equiv f(\mathbf{x}) + \sum_i k_i[g_i(\mathbf{x})]^2,$$

where the second term on the right is the penalty function. We find suitable values of the k_i by attempting to minimize $Z(\mathbf{x})$ in the directions $\nabla g_i(\mathbf{x})$. This gives

$$[\nabla Z(\mathbf{x})]'\nabla g_i(\mathbf{x}) = 0,$$

i.e.

$$\sum_j \left(\frac{\partial f}{\partial x_j} + 2\sum_i k_i g_i \frac{\partial g_i}{\partial x_j}\right)\frac{\partial g_r}{\partial x_j} = 0, \tag{5.14}$$

where r takes the same values as i; for example, $r = 1, \ldots, m$ if every constraint is an equation. Equations (5.14) are a set of simultaneous linear equations for the $k_i g_i$. If the partial derivatives in these equations are evaluated at the end of an iteration, then the resulting k_i are used for the next iteration.

Alternatively, having found the values of the $k_i g_i$ from equations (5.14), we may use

$$(k_i)_{new} = \frac{|g_i|}{\varepsilon_i}(k_i)_{old},$$

where the ε_i (>0) are the prescribed tolerances on the constraint functions $g_i(\mathbf{x})$. Normally, each k_i is increased at each iteration.

Equations (5.14) show that the violations g_i vary inversely as the corresponding k_i, and hence that $g_i \to 0$ as $k_i \to \infty$. This gives a theoretical justification for the general rule that it is advantageous to increase the values of the k_i as the optimal point is approached. In practice, however, it is not always advisable to determine the k_i by first solving equations (5.14) for the $k_i g_i$, since the computation may be time-consuming and, in spite of the above theory, equally good values of the k_i can often be found by trial and error. Finally, it is recommended that for greater accuracy the penalty function should be ignored for the last one or two iterations, the original constraints being used instead.

5.5 SEQUENTIAL UNCONSTRAINED MINIMIZATION TECHNIQUE (SUMT)

This method was proposed by Carroll[16] in 1961 and was developed by Fiacco and McCormick.[28–31] Carroll named it the 'Created Response Surface Technique'. The method replaces a constrained minimization problem by a sequence of unconstrained minimization problems. The basic idea is to attach different penalty functions to the given objective function in such a way that the optimal solutions of successive unconstrained problems approach the optimal solution of the given constrained problem.

Consider the constrained optimization problem:

$$\text{minimize} \quad z = f(\mathbf{x}),$$

subject to

$$g_i(\mathbf{x}) \geqslant 0.$$

This problem is replaced by the sequence of unconstrained problems:

$$\text{minimize } Z_k(\mathbf{x}, r_k) \equiv f(\mathbf{x}) + r_k \sum_i \frac{w_i}{g_i(\mathbf{x})} \quad (k = 1, 2, \ldots.), \tag{5.15}$$

where $r_k \geqslant 0$ for all k, and $w_i > 0$ for all i.

In problems (5.15), the w_i are weighting factors that remain fixed throughout the calculation, while the r_k are parameters that decrease from one iteration to the next. From a feasible initial point (see Section 5.1 for a method of locating such a point), and using any unconstrained optimization technique, problem (5.15) with $k = 1$ is solved. The minimizing point for this problem becomes the initial point for the next, with $k = 2$ and $r_2 < r_1$. This process is repeated until the desired accuracy is attained. If necessary, r_k is finally set equal to zero and a search is made for a minimum of the original objective function z, until a further decrease in z is prevented by a critical constraint or by limitations of accuracy.

It should be noted that the current point must remain feasible throughout the calculations. If a non-feasible point is reached at any time, then the calculations continue with a reduced step length from the last feasible point.

The convergence properties of the method have been thoroughly investigated by Fiacco and McCormick,[28] whose results we quote without proof. The method converges to give a unique minimum if the following conditions hold:

(i) The set $R_0 = \{\mathbf{x} : g_i(\mathbf{x}) > 0\}$ is non-empty.

(ii) The functions $f(\mathbf{x})$ and $-g_i(\mathbf{x})$ are convex and twice continuously differentiable.

(iii) For every finite value of K, the set $\{\mathbf{x} : f(\mathbf{x}) \leqslant K, \mathbf{x} \in R\}$ is bounded, where R is the closure of R_0.

(iv) For every $r_k > 0$, the modified objective function $Z_k(\mathbf{x}, r_k)$ is strictly convex.

Practical experience with the method has shown that it can also be applied successfully to many problems for which one or more of the above conditions is violated.

It is important to scale the variables x_j. Ideally, they should all be of the same order of magnitude and should change by similar amounts. The simplest way to deal with the weighting factors w_i is to set them all equal to unity. However, it is preferable to choose them in such a way that the initial values of the functions $w_i/g_i(\mathbf{x})$ are all of the same order of magnitude.

Fiacco and McCormick[29] suggest that the overall computational effort is closely associated with the choice of r_1. In view of the importance of this parameter, we shall discuss it at some length.

A suitable value for r_1 may be estimated as in Section 5.4, though trial and error may produce equally good results. Normally, $0{\cdot}5 \leqslant r_1 \leqslant 50$. Fiacco and McCormick recommend the formula

$$r_1 = \left\{\frac{[\nabla f(\mathbf{x}_1)]'\mathbf{H}_p^{-1}(\mathbf{x}_1)\nabla f(\mathbf{x}_1)}{[\nabla p(\mathbf{x}_1)]'\mathbf{H}_p^{-1}(\mathbf{x}_1)\nabla p(\mathbf{x}_1)}\right\}^{1/2}, \tag{5.16}$$

where $\mathbf{x}_1$ is the initial point and $\mathbf{H}_p(\mathbf{x}_1)$ is the Hessian matrix of

$$p(\mathbf{x}) \equiv \sum_i w_i/g_i(\mathbf{x}),$$

evaluated at $\mathbf{x} = \mathbf{x}_1$. Equation (5.16) is an estimate of the value of r_1 which minimizes

$$Z_1(\mathbf{x}_1, r_1) - Z_1(\mathbf{x}_m, r_1),$$

where $\mathbf{x} = \mathbf{x}_m$ is the point at which $Z_1(\mathbf{x}, r_1)$ is a minimum. This may be proved as follows.

To simplify the notation, let $Z(\mathbf{x}) \equiv Z_1(\mathbf{x}, r_1)$. Expand $Z(\mathbf{x})$ in a Taylor series about the point $\mathbf{x} = \mathbf{x}_m$, ignoring third- and higher-order terms:

$$Z(\mathbf{x}) = Z(\mathbf{x}_m) + (\mathbf{x} - \mathbf{x}_m)'\nabla Z(\mathbf{x}_m) + \tfrac{1}{2}(\mathbf{x} - \mathbf{x}_m)'\mathbf{H}_z(\mathbf{x}_m)(\mathbf{x} - \mathbf{x}_m), \tag{5.17}$$

where $\mathbf{H}_z(\mathbf{x}_m)$ is the Hessian matrix of $Z(\mathbf{x})$, evaluated at $\mathbf{x} = \mathbf{x}_m$. Since $\nabla Z(\mathbf{x}_m) = 0$, equation (5.17) gives

$$Z(\mathbf{x}) - Z(\mathbf{x}_m) = \tfrac{1}{2}(\mathbf{x} - \mathbf{x}_m)'\mathbf{H}_z(\mathbf{x}_m)(\mathbf{x} - \mathbf{x}_m). \tag{5.18}$$

Now interchange $\mathbf{x}$ and $\mathbf{x}_m$ in equation (5.17), giving

$$Z(\mathbf{x}_m) = Z(\mathbf{x}) + (\mathbf{x}_m - \mathbf{x})'\nabla Z(\mathbf{x}) + \tfrac{1}{2}(\mathbf{x}_m - \mathbf{x})'\mathbf{H}_z(\mathbf{x})(\mathbf{x}_m - \mathbf{x}). \tag{5.19}$$

Assuming that $\mathbf{H}_z(\mathbf{x}) = \mathbf{H}_z(\mathbf{x}_m)$, which is exact when $Z(\mathbf{x})$ is a quadratic function and is otherwise an approximation, equations (5.18) and (5.19) can be combined to give

$$Z(\mathbf{x}) - Z(\mathbf{x}_m) = -\tfrac{1}{2}(\mathbf{x}_m - \mathbf{x})'\nabla Z(\mathbf{x}) = \tfrac{1}{2}(\mathbf{x} - \mathbf{x}_m)'\mathbf{H}_z(\mathbf{x})(\mathbf{x} - \mathbf{x}_m). \tag{5.20}$$

Consistent with the assumptions already made, we regard (5.20) as a pair of identities in $\mathbf{x}$. From the second of these, we obtain

$$\nabla Z(\mathbf{x}) = \mathbf{H}_z(\mathbf{x})(\mathbf{x} - \mathbf{x}_m),$$

or

$$\mathbf{x} - \mathbf{x}_m = \mathbf{H}_z^{-1}(\mathbf{x})\nabla Z(\mathbf{x}).$$

Substituting for $\mathbf{x} - \mathbf{x}_m$ in the first identity of (5.20), we obtain

$$Z(\mathbf{x}) - Z(\mathbf{x}_m) = \tfrac{1}{2}[\nabla Z(\mathbf{x})]'\mathbf{H}_z^{-1}(\mathbf{x})\nabla Z(\mathbf{x}). \tag{5.21}$$

Thus, at the initial point $\mathbf{x} = \mathbf{x}_1$, we have

$$Z(\mathbf{x}_1) - Z(\mathbf{x}_m) = \tfrac{1}{2}[\nabla Z(\mathbf{x}_1)]'\mathbf{H}_z^{-1}(\mathbf{x}_1)\nabla Z(\mathbf{x}_1). \tag{5.22}$$

Now

$$Z(\mathbf{x}) \equiv f(\mathbf{x}) + r_1 p(\mathbf{x}),$$
$$\nabla Z(\mathbf{x}) \equiv \nabla f(\mathbf{x}) + r_1 \nabla p(\mathbf{x})$$

and

$$\mathbf{H}_z(\mathbf{x}) \equiv \mathbf{H}_f(\mathbf{x}) + r_1 \mathbf{H}_p(\mathbf{x}),$$

where $\mathbf{H}_f(\mathbf{x})$ and $\mathbf{H}_p(\mathbf{x})$ are the Hessian matrices of $f(\mathbf{x})$ and $p(\mathbf{x})$ respectively. Assuming that $\mathbf{H}_f(\mathbf{x}_1)$ is unimportant compared with $r_1\mathbf{H}_p(\mathbf{x}_1)$—this assumption is certainly valid if $\mathbf{x}_1$ is close to one or more of the constraint boundaries—equation (5.22) gives

$$Z(\mathbf{x}_1) - Z(\mathbf{x}_m) = \frac{1}{2r_1}[\nabla f(\mathbf{x}_1) + r_1 \nabla p(\mathbf{x}_1)]'\mathbf{H}_p^{-1}(\mathbf{x}_1)[\nabla f(\mathbf{x}_1) + r_1 \nabla p(\mathbf{x}_1)],$$

and by simple differentiation it is easy to verify that the value of r_1 which minimizes $Z(\mathbf{x}_1) - Z(\mathbf{x}_m)$ is given by equation (5.16).

It is assumed in equation (5.16) that $\nabla f(\mathbf{x}_1)$ and $\nabla p(\mathbf{x}_1)$ are non-zero and that $\mathbf{H}_p^{-1}(\mathbf{x}_1)$ exists. If these conditions are not all satisfied then some other method must be used to determine r_1.

Exercise

Show that the value of r_1 which minimizes $|\nabla Z(\mathbf{x}_1)|$ is

$$-[\nabla f(\mathbf{x}_1)]'\nabla p(\mathbf{x}_1)/|\nabla p(\mathbf{x}_1)|^2.$$

The value chosen for r_1 should be such that the gradient vectors of z and Z_1 at the initial point make an acute angle with each other; then the reduction in Z_1 during the first minimization virtually ensures a reduction in z. Any subsequent increase in z during the course of the calculations may be suppressed either by shortening the step length within an iteration or by proceeding to the next iteration.

The rate of decrease of the parameters r_k from one iteration to the next has little effect on the total amount of computation, for the faster the r_k are decreased the harder it is to reach the minimizing point in each iteration. A constant reduction factor is recommended for the r_k; a factor of 0·1 is often used.

Three modified versions of the basic formulation (5.15) have been used successfully. First, the functions $g_i(\mathbf{x})$ may be replaced by $[g_i(\mathbf{x})]^2$. Secondly, the summation may be taken over only those constraints that have been violated. Thirdly, equality constraints

$$g_i(\mathbf{x}) = 0 \qquad (i = m+1, \ldots, p)$$

may be included[30] by adding to the right-hand side of equation (5.15) the further term

$$r_k^{-1/2} \sum_{i=m+1}^{p} [g_i(\mathbf{x})]^2 \quad \text{or} \quad r_k^{-1/3} \sum_{i=m+1}^{p} [g_i(\mathbf{x})]^2,$$

depending on whether $g_i(\mathbf{x})$ or $[g_i(\mathbf{x})]^2$ is used for the inequality constraints. Obviously, $r_k \neq 0$ in this case and it is not possible to carry out the final search on the active constraint boundaries.

Example 5.1

Minimize

$$z = f(\mathbf{x}) \equiv (x_1 - 2)^3 + (x_2 - 3)^2,$$

subject to

$$\begin{aligned} x_1 + \ \ x_2 &\geqslant 4, \\ 2x_1 - 3x_2 &\geqslant 0. \end{aligned}$$

Solution

The constraint functions are

$$\begin{aligned} g_1(\mathbf{x}) &\equiv x_1 + x_2 - 4, \\ g_2(\mathbf{x}) &\equiv 2x_1 - 3x_2. \end{aligned}$$

We choose the feasible initial point $\mathbf{x}_1 = [4,1]$ and then choose w_1, w_2 such that

$$w_1/g_1(\mathbf{x}_1) = w_2/g_2(\mathbf{x}_1),$$

giving $w_1 = 1$, $w_2 = 5$. From equation (5.15), the modified objective functions are

$$\begin{aligned} Z_k(\mathbf{x}, r_k) &\equiv (x_1 - 2)^3 + (x_2 - 3)^2 + r_k\left(\frac{1}{x_1 + x_2 - 4} + \frac{5}{2x_1 - 3x_2}\right) \\ &\equiv f(\mathbf{x}) + r_k p(\mathbf{x}) \qquad (k = 1, 2, \ldots.). \end{aligned} \tag{5.23}$$

For numerical convenience we take $r_1 = 10$; this order of magnitude is indicated by equation (5.16).

To carry out the minimizations, we use the Newton–Raphson method (Section 4.3) with a linear search by Powell's quadratic interpolation method (Section 3.8). Thus we require the gradient vector $\mathbf{g}_k(\mathbf{x})$ and Hessian matrix $\mathbf{H}_k(\mathbf{x})$ of $Z_k(\mathbf{x}, r_k)$ with respect to $\mathbf{x}$. From equations (5.23), we find

$$\mathbf{g}_k(\mathbf{x}) \equiv \left[3(x_1 - 2)^2 - r_k\left\{\frac{1}{g_1{}^2(\mathbf{x})} + \frac{10}{g_2{}^2(\mathbf{x})}\right\},\ 2(x_2 - 3) - r_k\left\{\frac{1}{g_1{}^2(\mathbf{x})} - \frac{15}{g_2{}^2(\mathbf{x})}\right\}\right],$$

$$\mathbf{H}_k(\mathbf{x}) \equiv \begin{pmatrix} 6(x_1 - 2) + r_k\left\{\dfrac{2}{g_1{}^3(\mathbf{x})} + \dfrac{40}{g_2{}^3(\mathbf{x})}\right\}, & r_k\left\{\dfrac{2}{g_1{}^3(\mathbf{x})} - \dfrac{60}{g_2{}^3(\mathbf{x})}\right\} \\ r_k\left\{\dfrac{2}{g_1{}^3(\mathbf{x})} - \dfrac{60}{g_2{}^3(\mathbf{x})}\right\}, & 2 + r_k\left\{\dfrac{2}{g_1{}^3(\mathbf{x})} + \dfrac{90}{g_2{}^3(\mathbf{x})}\right\} \end{pmatrix}.$$

The results are summarized in Table 5.1; the values quoted refer to the beginning of each iteration. At the beginning of iteration 6, it becomes clear that the constraint $g_2(\mathbf{x}) \geqslant 0$ is active. Hence, instead of continuing with this iteration, a final search is made along the line $g_2(\mathbf{x}) = 0$, with $r_k = 0$, starting

Table 5.1 Solution of Example 5.1

Iteration Number k	1	2	3
x_1	4	3·9311	3·1167
x_2	1	1·2739	2·0187
$g_1(\mathbf{x})$	1	1·2050	1·1354
$g_2(\mathbf{x})$	5	4·0405	0·1773
$f(\mathbf{x})$	12	10·1808	2·3555
r_k	10	1	0·1
$r_k p(\mathbf{x})$	20	2·0674	2·9082
Z_k	32	12·2482	5·2637
$\mathbf{g}_k$	[−2, −8]	[9·8862, −3·2220]	[−28·148, 45·677]
$\mathbf{H}_k$	$\begin{pmatrix} 35·2 & 15·2 \\ 15·2 & 29·2 \end{pmatrix}$	$\begin{pmatrix} 13·5004 & -1·4894 \\ -1·4894 & 5·9556 \end{pmatrix}$	$\begin{pmatrix} 724·52 & -1076·38 \\ -1076·38 & 1616·92 \end{pmatrix}$
$\mathbf{H}_k^{-1}\mathbf{g}_k$	[0·0793, −0·3153]	[0·8144, −0·7448]	[0·2833, 0·2169]

Iteration Number k	4	5	6	Final Search	
				Start	End
x_1	2·7449	2·7672	2·7414	2·7414	2·7252
x_2	1·7340	1·7866	1·8224	1·8276	1·8168
$g_1(\mathbf{x})$	0·4789	0·5538	0·5638	0·5690	0·5420
$g_2(\mathbf{x})$	0·2878	0·1746	0·0156	0	0
$f(\mathbf{x})$	2·0161	1·9239	1·7943	1·7821	1·7814
r_k	0·01	0·001		0	
$r_k p(\mathbf{x})$	0·1946	0·0286			
Z_k	2·2107	1·9525			
$\mathbf{g}_k$	[0·4137, −0·7646]	[1·4345, −1·9380]			
$\mathbf{H}_k$	$\begin{pmatrix} 21·431 & -24·988 \\ -24·988 & 39·937 \end{pmatrix}$	$\begin{pmatrix} 12·130 & -11·261 \\ -11·261 & 18·920 \end{pmatrix}$			
$\mathbf{H}_k^{-1}\mathbf{g}_k$	[−0·01116, −0·02613]	[0·05164, −0·07151]			

at the point (2·7414, 1·8276). This simple linear search provides the final optimal solution:

$$\mathbf{x}^* = [2·7252, 1·8168]; \; z^* = 1·7814.$$

5.6 THE DFP METHOD WITH LINEAR CONSTRAINTS

The Davidon–Fletcher–Powell (DFP) method (Section 4.4) is an efficient general purpose technique for unconstrained optimization; it is therefore not surprising that attempts have been made to adapt it to solve constrained optimization problems. The DFP method can, of course, be used in conjunction with SUMT (Section 5.5), but a disadvantage of SUMT is that it takes no account of any special features of the constraints, e.g. their linearity.

The method we shall describe here is due in the first place to Davidon,[24] who suggested that active linear constraints could be included in his method by modifying the matrix $\mathbf{H}_k$ in such a way that these constraints are satisfied identically along the search direction. In other words, the matrix $\mathbf{H}_k$ is modified so that it becomes a suitable projection matrix. This idea was developed by Goldfarb,[40] Goldfarb and Lapidus,[41] and Fletcher.[33] It has been shown by Davies[25] that a similar method may be used when nonlinear constraints are present. The method may also be used when the complementary DFP formula (4.41) replaces the DFP formulae (4.17) to (4.19).

Consider first the case where every constraint is a linear equation, typically of the form

$$\mathbf{c}'\mathbf{x} = b. \tag{5.24}$$

The search direction in the absence of constraints is $\mathbf{d}_k = -\mathbf{H}_k\mathbf{g}_k$, where $\mathbf{H}_k$ is an $(n \times n)$ positive definite matrix, as explained in Section 4.4. When the constraint (5.24) is included, the search direction becomes

$$\mathbf{d}_{k1} = -\mathbf{H}_{k1}\mathbf{g}_k,$$

where

$$\mathbf{H}_{k1} = \mathbf{H}_k - \frac{\mathbf{H}_k\mathbf{c}\mathbf{c}'\mathbf{H}_k}{\mathbf{c}'\mathbf{H}_k\mathbf{c}}. \tag{5.25}$$

If the current point $\mathbf{x}_k$ satisfies the constraint (5.24) then any point $\mathbf{x}$ in the search direction $\mathbf{d}_{k1}$ from $\mathbf{x}_k$ will also satisfy this constraint, as required. For

$$\begin{aligned}\mathbf{c}'(\mathbf{x}_k + \lambda\mathbf{d}_{k1}) &= \mathbf{c}'\mathbf{x}_k - \lambda\mathbf{c}'\left(\mathbf{H}_k - \frac{\mathbf{H}_k\mathbf{c}\mathbf{c}'\mathbf{H}_k}{\mathbf{c}'\mathbf{H}_k\mathbf{c}}\right)\mathbf{g}_k \\ &= b - \lambda\left(\mathbf{c}'\mathbf{H}_k - \frac{\mathbf{c}'\mathbf{H}_k\mathbf{c}\mathbf{c}'\mathbf{H}_k}{\mathbf{c}'\mathbf{H}_k\mathbf{c}}\right)\mathbf{g}_k \\ &= b. \end{aligned} \tag{5.26}$$

When the matrix $\mathbf{H}_{k1}$ of equation (5.25) is updated in the usual way to give $\mathbf{H}_{k+1,1}$ —see equation (4.17)—any point in the new search direction $\mathbf{d}_{k+1,1} = -\mathbf{H}_{k+1,1}\mathbf{g}_{k+1}$ from $\mathbf{x}_{k+1}$ also satisfies the constraint (5.24), since $\mathbf{c}'\mathbf{H}_{k+1,1} = \mathbf{0}'$. Furthermore, it has been shown by Goldfarb[40] that the directions of search retain the important property of being mutually conjugate. As a consequence of this, the method will find the minimum of a quadratic function of n variables subject to q active linear constraints in, at most, $(n - q)$ iterations.

In the case where m linear equality constraints

$$\mathbf{c}_i'\mathbf{x} = b_i \tag{5.27}$$

are present, the above results are easily generalized. Following the convenient notation of Davies,[25] we form the $(n \times m)$ matrix $\mathbf{C}$ whose *columns* are the vectors $\mathbf{c}_i$. It is assumed, without loss of generality, that the columns of $\mathbf{C}$ are linearly independent. The constraints (5.27) then become

$$\mathbf{C}'\mathbf{x} = \mathbf{b},$$

where $\mathbf{b} = [b_1, ..., b_m]$; and the search direction is

$$\mathbf{d}_{km} = -\mathbf{H}_{km}\mathbf{g}_k,$$

where

$$\mathbf{H}_{km} = \mathbf{H}_k - \mathbf{H}_k\mathbf{C}(\mathbf{C}'\mathbf{H}_k\mathbf{C})^{-1}\mathbf{C}'\mathbf{H}_k. \tag{5.28}$$

The suffix m on $\mathbf{H}_{km}$ indicates that m active constraints are being taken into account. Again, if the current point $\mathbf{x}_k$ satisfies the constraints (5.27) then any point $\mathbf{x}$ in the search direction $\mathbf{d}_{km}$ from $\mathbf{x}_k$ will also satisfy these constraints. For

$$\begin{aligned}\mathbf{C}'(\mathbf{x}_k + \lambda\mathbf{d}_{km}) &= \mathbf{C}'\mathbf{x}_k - \lambda\mathbf{C}'[\mathbf{H}_k - \mathbf{H}_k\mathbf{C}(\mathbf{C}'\mathbf{H}_k\mathbf{C})^{-1}\mathbf{C}'\mathbf{H}_k]\mathbf{g}_k \\ &= \mathbf{b} - \lambda[\mathbf{C}'\mathbf{H}_k - \mathbf{C}'\mathbf{H}_k\mathbf{C}(\mathbf{C}'\mathbf{H}_k\mathbf{C})^{-1}\mathbf{C}'\mathbf{H}_k]\mathbf{g}_k \\ &= \mathbf{b}.\end{aligned}$$

As before, the direction of search remains feasible and the property of mutual conjugacy is retained when the matrix $\mathbf{H}_{km}$ of equation (5.28) is updated in the usual way.

The proof of equation (5.26) shows that the matrix $\mathbf{H}_{k1}$ of equation (5.25) satisfies

$$\mathbf{c}'\mathbf{H}_{k1} = \mathbf{0}'; \tag{5.29}$$

similarly, the matrix $\mathbf{H}_{km}$ of equation (5.28) satisfies

$$\mathbf{C}'\mathbf{H}_{km} = \mathbf{O}. \tag{5.30}$$

It follows that $\mathbf{H}_{km}$ may be generated from $\mathbf{H}_k$ by applying equation (5.25) to the m columns of $\mathbf{C}$ in turn. Furthermore, in equation (5.25), the second matrix on the right is of rank 1, and equation (5.29) is a linear relation between the rows of $\mathbf{H}_{k1}$ which does not hold between the rows of $\mathbf{H}_k$. Hence

$$\text{rank}(\mathbf{H}_{k1}) = \text{rank}(\mathbf{H}_k) - 1.$$

(We have assumed here the result that the rank of the sum of two matrices is not less than the difference between their ranks.) Similarly, in equation (5.28), the second matrix on the right is of rank m, and equation (5.30) represents m independent linear relations between the rows of $\mathbf{H}_{km}$, none of which holds between the rows of $\mathbf{H}_k$. Hence

$$\text{rank}(\mathbf{H}_{km}) = \text{rank}(\mathbf{H}_k) - m.$$

Thus the rank of $\mathbf{H}_k$ is reduced by one for each equality constraint.

Consider next the case where linear *inequality* constraints are present. Constraints of this type may be included in the DFP method in much the same way as equality constraints, as follows. Suppose that the inequality constraint

$$\mathbf{c}'\mathbf{x} \geqslant b \tag{5.31}$$

has to be satisfied, and that, at the current point,

$$\mathbf{c}'\mathbf{x}_k > b.$$

Then the constraint (5.31) is ignored until it becomes active; the point at which this happens may be determined by interpolation. If the next search direction is non-feasible, then the constraint

$$\mathbf{c}'\mathbf{x} = b \tag{5.32}$$

is added to the list of equality constraints before the matrix $\mathbf{H}_{km}$ of equation (5.28) is determined. Thus the inequality constraint (5.31) is either treated as an equality constraint or is ignored; and similarly for any other linear inequality constraint.

It may happen that at some later stage of the calculation the constraint (5.32) must be relaxed; it may become profitable to move the current point away from the constraint boundary. To allow for this possibility, we require an operation which is the inverse of (5.25), i.e. we have to find a matrix $\mathbf{H}_k$ such that the new search direction $-\mathbf{H}_k\mathbf{g}_k$ does not follow the constraint boundary (5.32). This is accomplished by adding a matrix of rank 1 to $\mathbf{H}_{k1}$:

$$\mathbf{H}_k = \mathbf{H}_{k1} + \frac{\mathbf{c}\mathbf{c}'}{\mathbf{c}^2}, \tag{5.33}$$

for if $\mathbf{x} = \mathbf{x}_k - \lambda\mathbf{H}_k\mathbf{g}_k$ then $\mathbf{c}'\mathbf{x} = b - \lambda\mathbf{c}'\mathbf{g}_k \neq b$ in this case.

We shall now show that equation (5.33) implies

$$\operatorname{rank}(\mathbf{H}_k) = \operatorname{rank}(\mathbf{H}_{k1}) + 1. \tag{5.34}$$

First, the linear relation between the rows of $\mathbf{H}_{k1}$, indicated by equation (5.29), does not hold between the rows of $\mathbf{H}_k$, for

$$\mathbf{c}'\mathbf{H}_k = \mathbf{c}'\mathbf{H}_{k1} + \frac{\mathbf{c}^2\mathbf{c}'}{\mathbf{c}^2} = \mathbf{c}' \neq \mathbf{0}'.$$

Secondly, suppose that a linear relation holds between the rows of $\mathbf{H}_k$ that does not hold between the rows of $\mathbf{H}_{k1}$. Then, for some vector $\mathbf{u} \neq \mathbf{0}$,

$$\mathbf{u}'\mathbf{H}_k = \mathbf{0}' \quad \text{and} \quad \mathbf{u}'\mathbf{H}_{k1} \neq \mathbf{0}'. \tag{5.35}$$

Hence, using equation (5.33), we have

$$\mathbf{u}'\mathbf{H}_k = \mathbf{u}'\mathbf{H}_{k1} + \frac{\mathbf{u}'\mathbf{c}\mathbf{c}'}{\mathbf{c}^2} = \mathbf{0}', \tag{5.36}$$

and so

$$\mathbf{u}'\mathbf{H}_{k1}\mathbf{c} + \frac{\mathbf{u}'\mathbf{c}\mathbf{c}^2}{\mathbf{c}^2} = 0. \tag{5.37}$$

But $\mathbf{H}_{k1}\mathbf{c} = \mathbf{0}$ and so $\mathbf{u}'\mathbf{c} = 0$, from equation (5.37). Thus equation (5.36) implies $\mathbf{u}'\mathbf{H}_{k1} = \mathbf{0}'$, which contradicts the second equation of (5.35). It follows that the transformation (5.33) makes the rank of $\mathbf{H}_k$ exactly one more than the rank of $\mathbf{H}_{k1}$, as stated by equation (5.34).

Equation (5.33) is used when $\mathbf{c}'\mathbf{x} = b$ is the only active constraint. More generally, suppose that q constraints

$$\mathbf{c}_i'\mathbf{x} \geqslant b_i \qquad (i = 1, \ldots, q) \tag{5.38}$$

are active and that one of them is to be relaxed. It is convenient to write equations (5.38) in matrix form:

$$\mathbf{C}'\mathbf{x} \geqslant \mathbf{b}.$$

Instead of equation (5.33), we now use

$$\mathbf{H}_{k,q-1} = \mathbf{H}_{kq} + \frac{\mathbf{P}_{q-1}\mathbf{c}_r\mathbf{c}_r'\mathbf{P}_{q-1}}{\mathbf{c}_r'\mathbf{P}_{q-1}\mathbf{c}_r}, \tag{5.39}$$

where $\mathbf{c}_r'\mathbf{x} = b_r$ is the constraint to be relaxed, and the $(n \times n)$ projection matrix $\mathbf{P}_{q-1}$ is defined by

$$\mathbf{P}_{q-1} = \mathbf{I} - \mathbf{C}_{q-1}(\mathbf{C}_{q-1}'\mathbf{C}_{q-1})^{-1}\mathbf{C}_{q-1}',$$

in which the $(n \times (q-1))$ matrix $\mathbf{C}_{q-1}$ is the matrix $\mathbf{C}$ with its rth column deleted. The validity of equation (5.39) depends on the property of $\mathbf{P}_{q-1}$ that $\mathbf{P}_{q-1}\mathbf{c}_r \neq 0$, while $\mathbf{P}_{q-1}\mathbf{c}_i = 0$, $i \neq r$ (Section 5.3). When more than one constraint is to be relaxed, equation (5.39) is applied repeatedly.

There remains the problem of deciding whether it is profitable to relax a constraint. Let the active constraints be

$$\mathbf{c}_i'\mathbf{x} \geqslant b_i \qquad (i = 1, \ldots, q)$$

or

$$\mathbf{C}'\mathbf{x} \geqslant \mathbf{b},$$

and let $\mathbf{g}_k$ be the current gradient direction. Then there exists a q-vector $\boldsymbol{\lambda}$ such that

$$\mathbf{g}_k = \mathbf{C}\boldsymbol{\lambda}$$

or

$$\mathbf{g}_k = \sum_{i=1}^{q} \lambda_i \mathbf{c}_i. \tag{5.40}$$

In fact,

$$\boldsymbol{\lambda} = (\mathbf{C}'\mathbf{C})^{-1}\mathbf{C}'\mathbf{g}_k. \tag{5.41}$$

The vector $\boldsymbol{\lambda}$ can be identified as the vector of Lagrange multipliers associated with the active constraints. This follows by comparing equations (2.45) and (5.40), ignoring the terms due to non-negativity restrictions in the former. At the optimal point, the Lagrange multipliers must satisfy the conditions of Theorem 2.1 (page 36); see also equation (1.46) *et seq*. In particular, Theorem 2.1 implies that for the present *minimizing* problem,

$$\boldsymbol{\lambda}^* \geqslant 0. \tag{5.42}$$

It is desirable to relax every constraint for which the corresponding component of λ is negative. In practice, however, in order to reduce the computational effort, only the constraint corresponding to the most negative component of λ is relaxed in each iteration.

The computational procedure for the application of the DFP method to a minimizing problem with linear constraints may be summarized briefly as follows:

1. Choose an initial feasible point.
2. Determine the search direction $\mathbf{d}_k = -\mathbf{H}_k\mathbf{g}_k$ as in equation (4.12) and find the components of λ from equation (5.41). If $\mathbf{d}_k = 0$ and $\lambda \geqslant 0$ then the optimal point has been reached; a normal stopping condition is inserted here (see step 9, page 111).
3. Relax the constraint corresponding to the component of λ which has the most negative value, using equation (5.33) or (5.39). Re-determine the search direction and check that it is feasible with respect to all the passive constraints; condition (5.42) ensures only that the direction $-\mathbf{g}_k$ is feasible. If the search direction is not feasible, modify the matrix $\mathbf{H}_k$ using equation (5.25) or (5.28), re-determine the search direction, etc., and continue until the search direction is feasible.
4. Carry out the linear search.
5. If a constraint boundary is reached during the initial or any subsequent linear search, move the current point to that boundary and modify the matrix $\mathbf{H}_k$ using equation (5.25) or (5.28). Return to step 2 if $\mathbf{d}_k = 0$. Otherwise, perform another linear search using the modified $\mathbf{H}_k$ and repeat this step until a minimum is found at a feasible point.
6. Update the matrix $\mathbf{H}_k$ in the usual way (see step 7, page 111). Return to step 2.

Further computational details of this method, with numerical examples, are given by Davies,[25] who also shows how the method can be extended to take account of nonlinear constraints. Essentially, any active nonlinear constraint hypersurface is approximated locally by a tangent hyperplane, and the theory for linear constraints is applied, together with a hemstitching procedure.

SUMMARY

Only in a few simple cases can constraints be eliminated by a change of variables. Normally, the procedure is to transform the given constrained problem into an unconstrained problem. In the method known as 'hemstitching', the constraints are ignored until one or more of them is violated; the current point then returns to the feasible region and the constraints are again ignored, etc. Several variants of this method have been used successfully in practice. The gradient projection method also ignores passive constraints, but moves the current point in a direction tangential to the set of active constraint boundaries, this direction being determined by the projection matrix $\mathbf{P}$

of equation (5.7). The idea of a penalty function is very simple and attractive: a high penalty is imposed on the value of the objective function if the constraint is violated. The major difficulty with this method lies in the choice of suitable parameters. A sequence of penalty functions is used in the method known as SUMT; this method has proved to be one of the most successful general purpose methods for dealing with constraints. Any of the above methods can be used in conjunction with any unconstrained optimization technique. In contrast, a special modification of the DFP method exists for dealing with linear constraints.

EXERCISES

1. Find suitable changes of variables to eliminate the following constraints.
(a) $x_1^2 - x_2^2 \leqslant 1$.
(b) $-1 \leqslant x_1^2 - x_2^2 \leqslant 1$.
(c) $x_1^2 + 2x_2^2 \leqslant a^2$.
(d) $x_1 x_2 x_3 \leqslant a^3$.
(e) $x_1^2 \leqslant 4a(x_2 + b)$, $x_1 \geqslant 0$, $x_2 \leqslant 0$, $(a > 0, b > 0)$.
2. Prove that the constraint

$$\sum_j x_j^2 = 1$$

may be eliminated by using either of the following transformations.
(a) $x_1 = \prod_{j=1}^{n-1} \sin y_j$; $x_j = \cos y_{j-1} \prod_{k=j}^{n-1} \sin y_k (j=2,\ldots,n-1)$; $x_n = \cos y_{n-1}$.
(b) $x_j = y_j/a \; (j=1,\ldots,n-1)$: $x_n = 1/a$;
where

$$a = \left(1 + \sum_{j=1}^{n-1} y_j^2\right)^{1/2}.$$

3. Find the direction of search from the point [2,2] when the gradient projection method is applied to the following problem:

$$\text{minimize} \quad z = x_1^2 + 3x_1x_2 + 5x_2^2 - 5x_1,$$

subject to

$$x_1^2 - 2x_2 \leqslant 0,$$
$$x_2^2 - 2x_1 \leqslant 0.$$

4. For the problem of Exercise 3, use the gradient projection method to find the return direction from the point [2,1].
5. Solve the following linear programming problem by the gradient projection method, starting from the point [0,2,2]:

$$\text{minimize} \quad z = x_1 + 2x_2 - 3x_3,$$

subject to

$$x_1 + x_2 + x_3 \geqslant 4,$$

$$4x_1 + 3x_2 + 2x_3 \leqslant 10,$$
$$x_1, x_2, x_3 \geqslant 0.$$

6. Let $P(\mathbf{x} = \mathbf{x}_k)$ be a non-feasible point close to the constraint boundary $g(\mathbf{x}) = 0$. As a first approximation, the return direction from P to the constraint boundary is assumed to be in the direction

$$\mathbf{r} = \pm \nabla g(\mathbf{x}_k),$$

according as the constraint is $g(\mathbf{x}) \geqslant$ or $\leqslant 0$. Use the first mean value theorem to prove that a better estimate of the return direction from P is

$$\left[\mathbf{I} - \frac{g(\mathbf{x}_k)}{|\nabla g(\mathbf{x}_k)|^2} \mathbf{H}_g(\mathbf{x}_k)\right] \mathbf{r},$$

where $\mathbf{H}_g(\mathbf{x})$ is the Hessian matrix of $g(\mathbf{x})$.

Use these formulae to find first and second approximations to the return direction from the point [3,3,3] to the boundary of the constraint

$$x_1^3 - x_2 x_3^2 + x_1 x_2 x_3 \geqslant 30.$$

7. Find the error in using the second approximation of Exercise 6 when the constraint is

$$x_1^2 + 4x_2^2 \geqslant 4$$

and the non-feasible point P is $[1, \frac{1}{2}]$.

8. The penalty function

$$P(x) \equiv \left(\frac{4x}{\pi} - 3\right)^{2M}$$

is used to represent the constraints in the following problem:

$$\text{minimize} \quad f(x) \equiv x \sin x \quad \left(\frac{\pi}{2} \leqslant x \leqslant \pi\right).$$

If $P(\mathbf{x})$ is not discarded as the optimal point is approached, find the error in the value obtained for x^* when (a) $M = 1$, (b) $M = 5$.

9. Solve the minimization problem of Exercise 3, using a penalty function of type (5.12). Take $\mathbf{x}_1 = [1,1]$ as the initial point.

10. Solve the following problem by the method of steepest descent with a fixed step length, combined with hemstitching:

$$\text{maximize} \quad f(\mathbf{x}) \equiv x_1 x_2 x_3,$$

subject to

$$10x_1^2 + x_2^2 + 4x_3^2 \leqslant 69,$$
$$x_1, x_2, x_3 \geqslant 0.$$

11. Solve the following problem by using any unconstrained optimization technique combined with hemstitching:

$$\text{minimize} \quad f(\mathbf{x}) \equiv x_1^2 + x_2^2 + x_3^2,$$

subject to

$$x_1 x_2 x_3 \geqslant 3,$$
$$x_1 + x_2 - x_3 \geqslant 3.$$

Take $\mathbf{x}_1 = [3,2,1]$ as the initial point.

12. What happens to the problem of Exercise 11 if the additional constraint

$$x_1 x_2 + x_2 x_3 + x_3 x_1 \geqslant 3$$

is included?

13. Use SUMT to solve the following problem:

$$\text{maximize} \quad f(\mathbf{x}) \equiv x_1^4 + x_2^2,$$

subject to

$$x_1 x_2 \leqslant 8,$$
$$x_1, x_2 \geqslant 0.$$

14. Use SUMT to solve the following problem:

$$\text{maximize} \quad f(\mathbf{x}) \equiv x_1^2 + x_1 x_2 - x_2^2,$$

subject to

$$2x_2 \geqslant x_1,$$
$$x_2 \leqslant 2x_1,$$
$$1 \leqslant x_1^2 + x_2^2 \leqslant 2.$$

Take $\mathbf{x}_1 = [1, 0{\cdot}75]$ as the initial point.

15. Solve the problem of Example 5.1 by the DFP method.

16. Solve the following problem by the DFP method:

$$\text{maximize} \quad f(\mathbf{x}) \equiv x_1^2 + 4x_1 x_2 + 7x_2^2,$$

subject to

$$-1 \leqslant x_1 + 2x_2 \leqslant 1,$$
$$-1 \leqslant 3x_1 - 4x_2 \leqslant 1.$$

Take $\mathbf{x}_1 = [0{\cdot}2, 0{\cdot}2]$ as the initial point.

CHAPTER 6

Dynamic Programming

6.1 INTRODUCTION

Dynamic programming is a powerful optimization technique which may be applied to many problems whose solution involves a multistage decision process. That is to say, the problem is divided into a number of stages or sub-problems, and is solved sequentially by taking a decision at each stage. The stages are described by a *stage variable*, which usually takes the values 1, 2, 3, ... Apart from the stage variable, there are two other types of variable, or parameter, in dynamic programming. *State variables* describe the conditions of the problem at any stage of the solution, e.g. the available resources in an allocation problem, and values of the *decision variables* are chosen so as to obtain an optimal solution of a sub-problem at each stage.

The choice of values for the decision variables at any stage determines an *optimal policy* for that stage; ultimately, an optimal policy for the complete problem is obtained. A decision variable regarded as a function of the state variables is called a *policy function*. The variables in dynamic programming may be discrete or continuous; the latter case arises, for example, when the number of stages in the problem is allowed to increase indefinitely. Then the stage variable, e.g. the time t, becomes continuous, giving a continuous decision process. Even when the stage variable is discrete, the state and decision variables within each stage may be continuous; the farmer's problem (Section 6.5) is of this type.

In contrast to many other optimization techniques, the dynamic programming method normally yields a global optimal solution. Unfortunately, the requisite computational effort increases very rapidly with the number of state variables. If k state variables are necessary at each stage, then a k-dimensional search is required in order to find the optimal policy for each stage. In practice, even with the aid of special techniques and large computers, problems for which k is greater than three or four cannot normally be solved. In this chapter, we shall consider problems which require only one state variable at each stage. Methods for dealing with two or more state variables are dealt with in specialized texts.[5,6,57]

Not every multistage decision problem can be solved by dynamic programming. The problem must possess the property that when the value of the decision variable has been chosen for any stage, the remaining sub-problem has precisely the same structure as the previous sub-problem. This allows us to write down a recurrence relation, based on the 'Principle of Optimality' (Bellman):[5]

> An optimal policy has the property that, whatever the initial state and initial decision, the remaining decisions must constitute an optimal policy with regard to the state resulting from the initial decision.

Although the principle of optimality appears to be nothing more than common-sense, its application to problem-solving is decidedly non-trivial. The reason for this is that dynamic programming can be applied to a wide variety of problems, and though it is usually obvious how the decision variables should be defined, the same is not always true of the state variables. Experience is the best guide! A feature common to all dynamic programming solutions is that the given problem is imbedded within a family of similar problems formed by generalizing the data of the given problem.

The remainder of this chapter is devoted to some typical applications of the dynamic programming technique.

6.2 THE ALLOCATION PROBLEM

The general form of the allocation problem is:

$$\text{maximize} \quad z = \sum_j f_j(x_j), \tag{6.1}$$

subject to

$$\sum_j g_j(x_j) \leqslant b, \tag{6.2}$$

$$\left.\begin{aligned} g_j(x_j) &\geqslant 0, \\ x_j &\geqslant 0. \end{aligned}\right\} \tag{6.3}$$

The reason for the name 'allocation problem' is that finding optimal values of the x_j is equivalent to allocating the 'resources' b among the 'activities' $g_j(x_j)$ in such a way that the objective function z is maximized.

In the dynamic programming solution, the constraint (6.2) is generalized to

$$\sum_j g_j(x_j) \leqslant \xi_n \leqslant b, \tag{6.4}$$

where ξ_n is a state variable. The effect of introducing ξ_n is to imbed the given problem in the class of problems formed by replacing the constant b by the state variable ξ_n.

Let $F_n(\xi_n)$ denote the maximum value of z in (6.1), subject to the constraints (6.3) and (6.4). Thus

$$F_n(\xi_n) = \max_{x_1,\dots,x_n} \left\{ f_n(x_n) + \sum_{j=1}^{n-1} f_j(x_j) \right\}. \tag{6.5}$$

The subscript n on F denotes an n-stage problem; $F_n(\xi_n)$ is called a *maximum-return function* or simply a *return function*. The stage variable is j and the decision variables are the x_j. We have to find $F_n(b)$; as a first step we relate the n-stage problem (6.5) to an $(n-1)$-stage problem.

Suppose that the maximization with respect to x_n is carried out in (6.5), with the result that

$$F_n(\xi_n) = f_n[\bar{x}_n(\xi_n)] + \max_{x_1,\dots,x_{n-1}} \left\{ \sum_{j=1}^{n-1} f_j(x_j) \right\}, \tag{6.6}$$

where $x_1,\dots,x_{n-1}$ in the remaining maximization satisfy the constraints (6.3), together with

$$\sum_{j=1}^{n-1} g_j(x_j) \leqslant \xi_n - g_n[\bar{x}_n(\xi_n)], \tag{6.7}$$

the latter being the particular form of (6.4) that occurs after $\bar{x}_n$ has been determined. Equation (6.6) relates the n-stage problem to a similar $(n-1)$-stage problem, and is merely a statement of the principle of optimality. It may be written in the equivalent forms

$$F_n(\xi_n) = f_n[\bar{x}_n(\xi_n)] + F_{n-1}\{\xi_n - g_n[\bar{x}_n(\xi_n)]\} \tag{6.8}$$

and

$$F_n(\xi_n) = \max_{x_n} \{ f_n(x_n) + F_{n-1}[\xi_n - g_n(x_n)]\}, \tag{6.9}$$

where $x_n \geqslant 0$ and

$$g_n(x_n) \leqslant \xi_n.$$

If we write

$$\xi_{n-1} = \xi_n - g_n(x_n),$$

then equation (6.9) becomes

$$F_n(\xi_n) = \max_{x_n} \{ f_n(x_n) + F_{n-1}(\xi_{n-1})\}. \tag{6.10}$$

Also, from (6.6), (6.7) and (6.8), the solution of the remaining $(n-1)$-stage problem becomes

$$F_{n-1}(\xi_{n-1}) = \max_{x_1,\dots,x_{n-1}} \left\{ \sum_{j=1}^{n-1} f_j(x_j) \right\}, \tag{6.11}$$

where $x_1,\dots,x_{n-1}$ satisfy the constraints (6.3), together with

$$\sum_{j=1}^{n-1} g_j(x_j) \leqslant \xi_{n-1}. \tag{6.12}$$

The $(n-1)$-stage problem is now in precisely the same form as the original n-stage problem: cf. (6.4) with (6.12) and (6.5) with (6.11). If the $(n-1)$-stage problem can be solved, then, using equation (6.10), the n-stage problem can be solved.

Proceeding in the same way, the solution of the $(n-1)$-stage problem can be related to the solution of an $(n-2)$-stage problem. In general, we obtain the recurrence relation

$$F_k(\xi_k) = \max_{x_k} \{ f_k(x_k) + F_{k-1}(\xi_{k-1}) \}, \quad (k = 1, \ldots, n), \tag{6.13}$$

where

$$\xi_{k-1} = \xi_k - g_k(x_k) \qquad (k = 1, \ldots, n), \tag{6.14}$$

and x_k is constrained by (6.3) together with

$$g_k(x_k) \leqslant \xi_k. \tag{6.15}$$

When $k = 1$ in equation (6.13), we take $F_0(\xi_0) = 0$.

The functions $F_k(\xi_k)$ of equation (6.13) must be evaluated in the order $F_1(\xi_1), \ldots, F_n(\xi_n)$ by maximizations over $x_1, \ldots, x_n$, respectively; we cannot evaluate $F_k(\xi_k)$ in (6.13) until we know the values of $F_{k-1}(\xi_{k-1})$ for every admissible value of ξ_{k-1}. The first function to be evaluated in this *forward solution* is

$$F_1(\xi_1) = \max_{x_1} f_1(x_1), \tag{6.16}$$

where x_1 must satisfy $x_1 \geqslant 0$ and

$$g_1(x_1) \leqslant \xi_1. \tag{6.17}$$

The function $F_1(\xi_1)$ of equation (6.16) is evaluated for every admissible value of ξ_1 in the closed interval $[0, b]$. Thus the search for the maximum of $f_1(x_1)$ with respect to x_1 becomes *simpler* as more restrictions are placed on the possible values that x_1 can take; and similarly for the other variables. In this respect, dynamic programming differs from most other optimization techniques. Restricting the variables to be integers, for example, normally increases the difficulty of an optimization technique.

Having evaluated $F_1(\xi_1)$ for every admissible value of ξ_1 in $[0, b]$, it is possible to evaluate $F_2(\xi_2)$ for every admissible value of ξ_2 in the same closed interval:

$$F_2(\xi_2) = \max_{x_2} \{ f_2(x_2) + F_1(\xi_1) \}, \tag{6.18}$$

where x_2 must satisfy $x_2 \geqslant 0$ and

$$g_2(x_2) \leqslant \xi_2,$$

with

$$\xi_1 = \xi_2 - g_2(x_2).$$

Similarly, $F_3(\xi_3), F_4(\xi_4), \ldots$ are evaluated, and ultimately we obtain

$$z^* = F_n(b). \tag{6.19}$$

The optimal values of the x_j are then immediately available from the previously tabulated results. We obtain, in turn,

$$x_n^* = \bar{x}_n(b),\ \xi_{n-1}^* = b - g_n(x_n^*),\ x_{n-1}^* = \bar{x}_{n-1}(\xi_{n-1}^*),$$
$$\xi_{n-2}^* = \xi_{n-1}^* - g_{n-1}(x_{n-1}^*),\ x_{n-2}^* = \bar{x}_{n-2}(\xi_{n-2}^*), \ldots, x_1^* = \bar{x}_1(\xi_1^*).$$

Example 6.1

Maximize

$$z = x_1^2 - 2x_1 + x_2^2 + 3x_2 + x_3^2 + 4x_3,$$

subject to

$$x_1^2 + 2x_2^2 + 3x_3^2 \leqslant 17,$$
$$x_1, x_2, x_3 \quad \textit{non-negative integers}. \tag{6.20}$$

Solution

From (6.16) and (6.17), we define

$$F_1(\xi_1) = \max_{x_1} \{x_1^2 - 2x_1\},$$

Table 6.1 Values of $F_1(\xi_1)$ and $\bar{x}_1(\xi_1)$ in Example 6.1

ξ_1	x_1	$F_1(\xi_1)$	$\bar{x}_1(\xi_1)$
0	0	0	0
1	0,1	0	0
2	0,1	0	0
3	0,1	0	0
4	0,1,2	0	0,2
5	0,1,2	0	0,2
6	0,1,2	0	0,2
7	0,1,2	0	0,2
8	0,1,2	0	0,2
9	0,1,2,3	3	3
10	0,1,2,3	3	3
11	0,1,2,3	3	3
12	0,1,2,3	3	3
13	0,1,2,3	3	3
14	0,1,2,3	3	3
15	0,1,2,3	3	3
16	0,1,2,3,4	8	4
17	0,1,2,3,4	8	4

where

$$0 \leqslant x_1^2 \leqslant \xi_1 \leqslant 17.$$

Because of condition (6.20), ξ_1 takes the values 0, 1,...., 17, and x_1 takes the values 0, 1, 2, 3, 4. Hence we obtain Table 6.1.

Next, we define

$$F_2(\xi_2) = \max_{x_2} \{x_2^2 + 3x_2 + F_1(\xi_1)\}, \tag{6.21}$$

as in equation (6.18) (or equation (6.13) with $k = 2$). From (6.14) and (6.15), we have

$$\xi_1 = \xi_2 - 2x_2^2$$

and

$$0 \leqslant 2x_2^2 \leqslant \xi_2 \leqslant 17,$$

so that ξ_2 takes the values 0, 1,...., 17, and x_2 takes the values 0, 1, 2. The values of $F_1(\xi_1)$ in equation (6.21) are taken from Table 6.1. Hence we obtain Table 6.2.

Finally, we define

Table 6.2 Values of $F_2(\xi_2)$ and $\bar{x}_2(\xi_2)$ in Example 6.1

ξ_2	x_2	ξ_1	$F_2(\xi_2)$	$\bar{x}_2(\xi_2)$
0	0	0	0	0
1	0	1	0	0
2	0,1	2,0	4	1
3	0,1	3,1	4	1
4	0,1	4,2	4	1
5	0,1	5,3	4	1
6	0,1	6,4	4	1
7	0,1	7,5	4	1
8	0,1,2	8,6,0	10	2
9	0,1,2	9,7,1	10	2
10	0,1,2	10,8,2	10	2
11	0,1,2	11,9,3	10	2
12	0,1,2	12,10,4	10	2
13	0,1,2	13,11,5	10	2
14	0,1,2	14,12,6	10	2
15	0,1,2	15,13,7	10	2
16	0,1,2	16,14,8	10	2
17	0,1,2	17,15,9	13	2

$$F_3(\xi_3) = \max_{x_3} \{x_3^2 + 4x_3 + F_2(\xi_2)\}, \tag{6.22}$$

using equation (6.13) with $k = 3$. From (6.14) and (6.15), we have

$$\xi_2 = \xi_3 - 3x_3^2 \tag{6.23}$$

and

$$0 \leqslant 3x_3^{\,2} \leqslant \xi_3 \leqslant 17. \tag{6.24}$$

However, we only need the value of $F_3(17)$ from equation (6.22), since this is the optimal value of z, by equation (6.19). From (6.24),

$$x_3 = 0, 1, 2,$$

and from equation (6.23), correspondingly,

$$\xi_2 = 17, 14, 5.$$

Hence from equation (6.22) and Table 6.2,

$$z^* = F_3(17) = \max\{13, 15, 16\} = 16. \tag{6.25}$$

Having obtained the optimal value z^* of z, we can find x_1^*, x_2^*, x_3^* in reverse order. Using the last of equations (6.25), we see that

$$x_3^* = \bar{x}_3(17) = 2,$$

corresponding to $\xi_2^* = 5$. Then, from Table 6.2,

$$x_2^* = \bar{x}_2(5) = 1,$$

corresponding to $\xi_1^* = 3$, and from Table 6.1,

$$x_1^* = \bar{x}_1(3) = 0.$$

The optimal solution is therefore

$$\mathbf{x}^* = [0, 1, 2]; \; z^* = 16.$$

6.3 ORIENTED NETWORKS

Figure 6.1 represents a network which consists of links of given lengths, joined together at nodes. The 'length' of a link may represent the distance

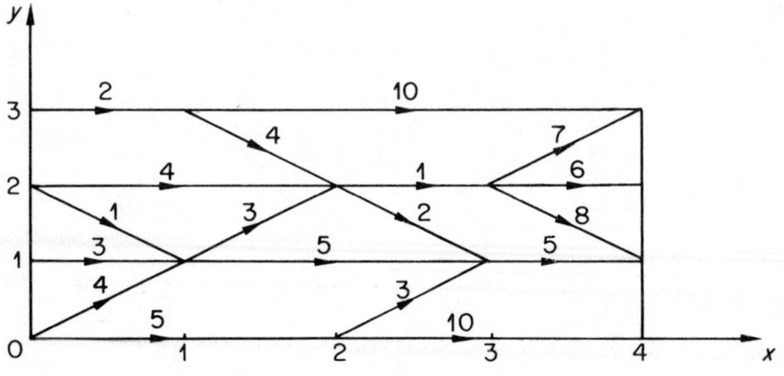

Figure 6.1 Oriented network

between its terminal nodes, the time taken to perform a task, the cost of transportation or communication between the nodes, etc. The network is said to be *oriented* because the admissible paths through it, indicated by the arrows, are all in the same general direction from left to right. Because of this property, it is convenient to use a coordinate system $O(x, y)$, as shown.

We shall find the path, or paths, of minimum length between the lines $x = 0$ and $x = 4$ by two almost identical methods—a forward solution and a backward solution. Along a minimal path, the notation $y(x)$ is used to denote the y-coordinate of the node at the point (x, y). Note that y is not necessarily a single-valued function of x: there may be more than one minimal path.

(*a*) *Forward solution*

The solution is divided into four stages, corresponding to the four unit intervals on the x-axis. Following the notation of Section 6.2, let $y(k) = \xi_k$, where k is the stage variable, and define the k-stage return function $F_k(\xi_k)$ as the length of the minimal path from the initial line $x = 0$ to the node (k, ξ_k).

The node (k, ξ_k) represents the state of the network at stage k; thus ξ_k is the state variable. A transformation from one state to the next involves the choice of a link joining the current node to a node at the next stage. The decision variable is therefore y or, equivalently, the policy function is $y(k)$.

Stage 1. $$F_1(1) = \min\{4, 3, 1\} = 1,$$
since the path of minimum length from $x = 0$ to $(1, 1)$ starts at $(0, 2)$ and is of length 1. This path is shown by a double arrow in Figure 6.2 and its length is circled. A similar notation is used for all minimal paths. Next,

$$F_1(3) = 2,$$

since there is only one possible path from $x = 0$ to $(1, 3)$. [Note that $F_1(0)$ and $F_1(2)$ do not exist for this network.]

The remaining stages can be followed by referring to Figures 6.1 and 6.2.

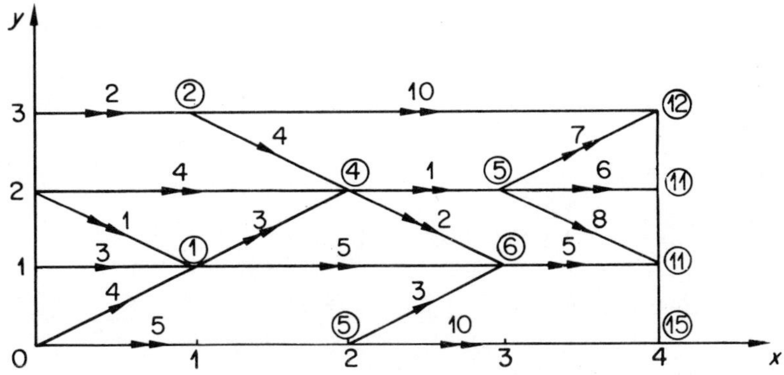

Figure 6.2 Forward solution for oriented network

Stage 2. $F_2(0) = 5,$
$F_2(2) = \min\{3 + F_1(1), 4, 4 + F_1(3)\} = \min\{4,4,6\} = 4.$
Stage 3. $F_3(1) = \min\{3 + F_2(0), 5 + F_1(1), 2 + F_2(2)\} = \min\{8,6,6\} = 6,$
$F_3(2) = 1 + F_2(2) = 5.$
Stage 4. $F_4(0) = 10 + F_2(0) = 15,$
$F_4(1) = \min\{5 + F_3(1), 8 + F_3(2)\} = \min\{11,13\} = 11,$
$F_4(2) = 6 + F_3(2) = 11,$
$F_4(3) = \min\{7 + F_3(2), 10 + F_1(2)\} = \min\{12,12\} = 12.$

The paths of minimum length are given by $F_4(1) = F_4(2) = 11$; they can be traced out by following the double arrows backwards from the terminal nodes (4, 1) and (4, 2). We find that there are five minimal paths, of which three terminate at the node (4, 1) and two at the node (4, 2). Every minimal path starts at the node (0, 2).

(*b*) *Backward solution*

As in the forward solution, suppose that $y(k) = \xi_k$. Let the k-stage return function $B_k(\xi_{4-k})$ denote the length of the minimal path from the node $(4 - k, \xi_{4-k})$ to the terminal line $x = 4$. The following steps can be followed by referring to Figures 6.1 and 6.3. Note that a backward solution becomes a forward solution if the diagram is turned upside-down.

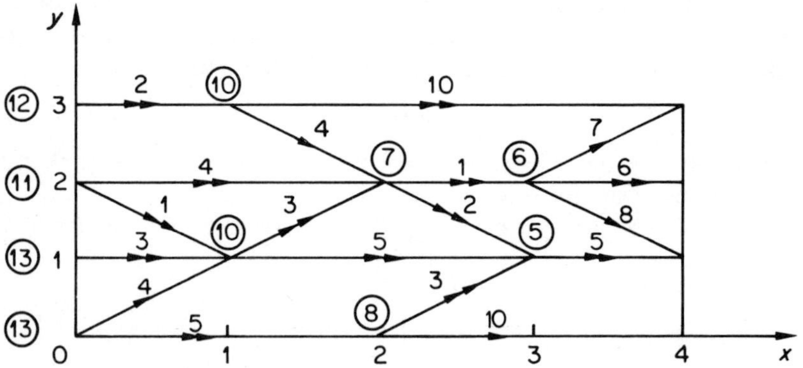

Figure 6.3 Backward solution for oriented network

Stage 1. $B_1(1) = 5,$
$B_1(2) = \min\{8,6,7\} = 6.$
Stage 2. $B_2(0) = \min\{3 + B_1(1), 10\} = \min\{8,10\} = 8,$
$B_2(2) = \min\{2 + B_1(1), 1 + B_1(2)\} = \min\{7,7\} = 7.$
Stage 3. $B_3(1) = \min\{5 + B_1(1), 3 + B_2(2)\} = \min\{10,10\} = 10,$
$B_3(3) = \min\{4 + B_2(2), 10\} = \min\{11,10\} = 10.$
Stage 4. $B_4(0) = \min\{5 + B_2(0), 4 + B_1(1)\} = \min\{13,14\} = 13,$
$B_4(1) = 3 + B_3(1) = 13,$

$$B_4(2) = \min\{1 + B_3(1), 4 + B_2(2)\} = \min\{11, 11\} = 11,$$
$$B_4(3) = 2 + B_3(3) = 12.$$

The paths of minimum length are given by $B_4(2) = 11$; they can be traced out by following the double arrows forwards from the initial node (0,2).

A forward solution is shorter than a backward solution when an initial node is specified, since only those paths that include the given node need be considered. When both an initial and terminal node are specified, either a forward or a backward solution is appropriate. To investigate the effects of a change in an initial node, a backward solution is obviously more suitable than a forward solution (the latter would have to be re-worked), and conversely for a change in a terminal node.

6.4 NON-ORIENTED NETWORKS

Consider a network of n nodes, numbered $1, 2, \ldots, n$, in which any pair of nodes may be linked and, in any link, the admissible direction of flow may be in one or both directions.

Let S_{I}, S_{T} be the non-empty sets of initial and terminal nodes, respectively, B_p the length of the minimal path from node p to S_{T}, and B_{IT} the length of the minimal path from S_{I} to S_{T}. Then

$$B_{\mathrm{IT}} = \min_{p \in S_{\mathrm{I}}} B_p.$$

If there is a direct link between node p and node q let l_{pq} be its length; otherwise let $l_{pq} = M$, a large positive value which need not be specified. Then the principle of optimality states that

$$B_p = \min_{q \neq p} \{l_{pq} + B_q\}, \tag{6.26}$$

with $B_q = 0$ when $q \in S_{\mathrm{T}}$. In equation (6.26), p is the stage or state variable (these variables coincide in the present problem) and q is the decision variable. Unfortunately, B_q $(q \notin S_{\mathrm{T}})$ is not known at the beginning of the calculation, so that equation (6.26) cannot be used as it stands: the unknown function appears on both sides of the equation.

To overcome this difficulty, we use an *approximation in function space*, i.e. an iterative algorithm for the generation of the return function B_p, starting from an initial guess of B_p as a function of the state variables. Equation (6.26) is replaced by the iterative formula

$$B_p^{k+1} = \min_{q \neq p} \{l_{pq} + B_q^{k}\} \quad (k = 0, 1, \ldots), \tag{6.27}$$

where B_q^0 is an initial approximation for B_q as a function of q. Probably the simplest approximation is to let B_q^0 be the minimum of the direct path lengths from node q to S_{T}; if there is no such direct path then we take $B_q^0 = M$. That is,

$$B_q^0 = \min_{r \in S_{\mathrm{T}}} l_{qr}. \tag{6.28}$$

Obviously, $B_q^0 = 0$ when $q \in S_T$.

***Theorem* 6.1**

The approximation in function space represented by equations (6.27) *and* (6.28) *ensures that* B_q^k *converges to the return function* B_q *in a finite number of iterations.*

Proof

Consider all the paths that begin at node p, end in S_T and have at most t links. When $t = 1$, the shortest of all these paths is of length B_p^0, by equation (6.28). When $t = 2$, the minimal path is of length B_p^1, using equation (6.27) with $k = 0$. In general, by applying equation (6.27) successively for $k = 1, \ldots, t-2$, we find that of all the paths consisting of not more than t links, the shortest is of length B_p^{t-1}.

Now a minimal path passes through any given node at most once, assuming that all lengths are positive. Hence there are not more than $(n-1)$ links in the overall minimal path from node p to S_T, and the length B_p of this path is therefore B_p^{n-2}, which proves the theorem.

Consider the special case of a single initial node and a single terminal node. For given values of p and k, the evaluation of B_p^{k+1} in equation (6.27) requires the comparison of at most $(n-1)$ values $(q = 1, \ldots, n;\ q \neq p)$. The B_p^{k+1} are evaluated for $p = 1, \ldots, n-1$, giving a total of at most $(n-1)^2$ comparisons for each value of k; but by Theorem 6.1, the only B_p^{k+1} that need be evaluated are $B_p^1, \ldots, B_p^{n-2}$, and so the total number of comparisons is at most $(n-1)^2$ $(n-2)$. The efficiency of the method may be measured by comparing this number with $(n-2)!e$, the latter being an approximation for the total number of loopless paths joining node 1 to node n. [A loopless path joining node 1 to node n passes through 0 or 1 or .... or $(n-2)$ other nodes, without repetition. Hence the total number of such paths is

$$
\begin{aligned}
&1 + (n-2) + (n-2)(n-3) + \ldots + (n-2)! \\
&= (n-2)!\left\{1 + \frac{1}{1!} + \frac{1}{2!} + \ldots + \frac{1}{(n-2)!}\right\} \\
&\doteqdot (n-2)!e.]
\end{aligned}
$$

Example 6.2

Find the path of minimum length from node 1 *to node* 7 *in the network of Figure* 6.4.

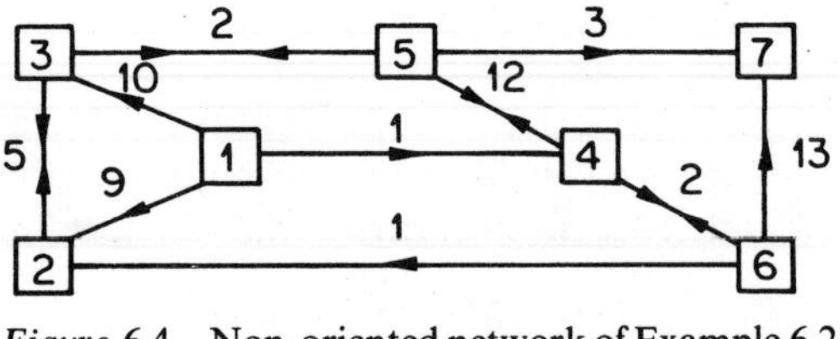

Figure 6.4 Non-oriented network of Example 6.2

Solution

All lengths are taken as positive. It is convenient to construct a matrix **L** whose elements are the lengths l_{pq} ($p = 1,....,7$, $q = 1,....,7$):

$$\mathbf{L} = \begin{pmatrix} 0 & 9 & 10 & 1 & M & M & M \\ M & 0 & 5 & M & M & M & M \\ M & 5 & 0 & M & 2 & M & M \\ M & M & M & 0 & 12 & 2 & M \\ M & M & 2 & 12 & 0 & M & 3 \\ M & 1 & M & 2 & M & 0 & 13 \\ M & M & M & M & M & M & 0 \end{pmatrix}$$

Next, we construct Table 6.3 row by row, giving the values of B_p^k. The values of B_p^0 ($p = 1,....,7$) appear in the last column of the matrix **L**; $B_7^k = 0$ for all k; and the remaining B_p^k are determined from equation (6.27) as follows.

Setting $k = 0$ in equation (6.27) gives

$$B_p^1 = \min_{q \neq p} \{l_{pq} + B_q^0\}.$$

Table 6.3 Values of B_p^k and q^* in Example 6.2

p	1	2	3	4	5	6	7
B_p^0	M	M	M	M	3	13	0
B_p^1	M	M	5	15	3	13	0
B_p^2	15	10	5	15	3	13	0
B_p^3	15	10	5	15	3	11	0
B_p^4	15	10	5	13	3	11	0
B_p^5	14	10	5	13	3	11	0
q^*	4	3	5	6	7	2	

Hence

$$B_1^1 = \min_{q \neq 1} \{l_{1q} + B_q^0\} = \min\{9+M, 10+M, 1+M, M+3, M+13, M\} = M,$$

$$B_2^1 = \min_{q \neq 2} \{l_{2q} + B_q^0\} = \min\{M+M, 5+M, M+M, M+3, M+13, M\} = M,$$

....

$$B_6^1 = \min_{q \neq 6} \{l_{6q} + B_q^0\} = \min\{M+M, 1+M, M+M, 2+M, M+3, 13\} = 13.$$

Setting $k = 1$ in equation (6.27) gives

$$B_p^2 = \min_{q \neq p} \{l_{pq} + B_q^1\}.$$

Hence

$$B_1^2 = \min_{q \neq 1} \{l_{1q} + B_q^1\} = \min\{9+M, 10+5, 1+15, M+3, M+13, M\} = 15,$$

$$B_2^2 = \min_{q \neq 2} \{l_{2q} + B_q^1\} = \min\{M+M, 5+5, M+15, M+3, M+13, M\} = 10,$$

and so on. Comparing the B_p^5 with the B_p^4, we see that only B_1^4 has changed, and hence there are no further changes in the values of the B_p^k. The minimal path is of length $B_1^5 = 14$. With each value of B_p^5 $(p = 1, \ldots, 6)$ is associated a minimal path from node p to node q^*, where q^* is the minimizing value of q in equation (6.27). The values of q^* appear in the last row of Table 6.3. Starting from node 1, we easily find the optimal path to be

$$1 \to 4 \to 6 \to 2 \to 3 \to 5 \to 7.$$

The method of dynamic programming can be used to find the second, third, ..., nth shortest path through a network. We shall consider the problem of finding the second shortest path; similar though more complex equations can be derived for the third, ..., nth shortest path.

Let C_p be the length of the second shortest path from node p to S_T. If node q follows node p in this path, then the path from node q to S_T is either the shortest path or the second shortest path. This leads to the following set of equations for the C_p.

Let mins(X_q) denote the second smallest value of X_q $(q = 1, \ldots, n)$. Then

$$C_p = \min\{\min_{q \neq p}(l_{pq} + C_q), \operatorname{mins}_{q \neq p}(l_{pq} + B_q)\}, \quad p \notin S_T, \tag{6.29}$$

where B_q is given by equation (6.26), i.e. by the solution of the shortest path problem, and C_p is undefined when $p \in S_T$ (the second shortest path cannot be of length zero).

Note. (i) In determining $\min_{q \neq p}(l_{pq} + C_q)$, we must exclude all q for which $q \in S_T$. However, in determining $\operatorname{mins}_{q \neq p}(l_{pq} + B_q)$, the values of q for which $q \in S_T$ are included.

(ii) Equation (6.29) for the length of the second shortest path does not exclude the possibility of a path being retraced. For example, the second shortest path from node 5 to node 7 in the network of Figure 6.4 is of length 7, given by

$$5 \to 3 \to 5 \to 7.$$

Example 6.3

Find the second shortest path from node 1 *to node* 7 *in the network of Figure* 6.4.

Solution

Making an approximation in function space—see equation (6.27)—we replace equation (6.29) by

$$C_p^{k+1} = \min\{\min_{q \neq p}(l_{pq} + C_q^k), \operatorname{mins}_{q \neq p}(l_{pq} + B_q)\}, \quad p \neq 7, \quad (k = 0, 1, \ldots), \tag{6.30}$$

where we take $C_q^0 = M$ $(q = 1, \ldots, 6)$. The proof that C_p^k converges to C_p in a finite number of iterations is closely similar to that of Theorem 6.1.

In equation (6.30), the l_{pq} and B_q are taken respectively from the matrix **L** and the B_p^5 row of Table 6.3. Thus $\min_{q \neq p}(l_{pq} + B_q)$ can be found for $p = 1,\ldots,6$; these values appear in the second row of Table 6.4, labelled *mins*. For $k = 0,1,2,\ldots$, the *mins* values are compared with the values of $\min_{q \neq p}(l_{pq} + C_q^k)$, labelled min_k in Table 6.4, to give the values of C_p^{k+1}

Table 6.4 Values of C_p^k in Example 6.3

p	1	2	3	4	5	6
mins	15	M	15	15	7	13
C_p^0	M	M	M	M	M	M
min_0	M	M	M	M	M	M
C_p^1	15	M	15	15	7	13
min_1	16	20	9	15	17	17
C_p^2	15	20	9	15	7	13
min_2	16	14	9	15	11	17
C_p^3	15	14	9	15	7	13
min_3	16	14	9	15	11	15
C_p^4	15	14	9	15	7	13

Since $C_p^4 = C_p^3$ for all values of p, there are no further changes in the values of the C_p^k. The second shortest path from node 1 to node 7 is therefore of length $C_1^3 = 15$. Also, we see that this value comes from the *mins* term of equation (6.30), for which the corresponding value of q^* is 3. Thus the second shortest path consists of the link joining node 1 to node 3, followed by the shortest path from node 3 to node 7, i.e.

$$1 \to 3 \to 5 \to 7.$$

6.5 THE FARMER'S PROBLEM

In a certain year, a farmer grows p tons of potatoes, of which some are used as seed for the following year and the remainder sold. The yield from each ton of seed potatoes is $m(>1)$ tons and the selling price of x tons of potatoes is $f(x)$. The variable x may be discrete or continuous and no restriction is placed on the function $f(x)$. In succeeding years, the farmer again keeps some potatoes for seed and sells the rest. Assuming that he sells the whole crop at the end of the nth year, what is his optimal selling policy if he wishes to maximize his total income?

This is a multistage problem with clearly defined stages. A backward solution is appropriate, since the policy for the final year, and hence the final 1-stage return function, is known. Thus we can determine the policy for the penultimate year, and so on. On the other hand, in a forward solution, both the initial policy and the initial 1-stage return function are unknown.

Let k be the time in years. Define the k-stage return function $B_k(\xi_{n-k})$ as the income that the farmer can earn by starting with ξ_{n-k} tons of potatoes and following an optimal policy for k years. The notation is shown in Figure 6.5. In order to clarify the timing, we assume that the state variables ξ_k refer to January in any year and that the decision variables x_k refer to December.

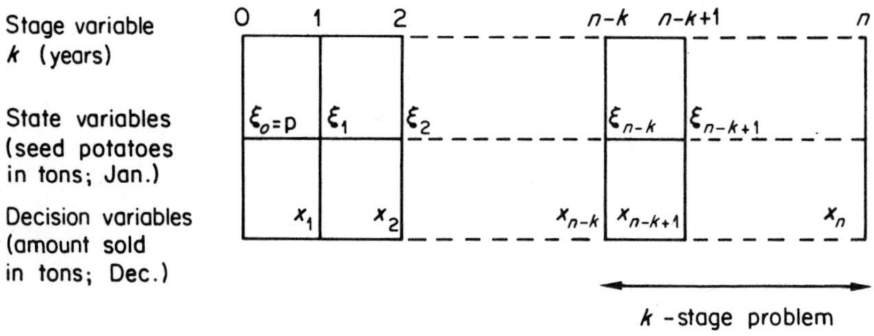

Figure 6.5 Notation for backward solution

The farmer begins by growing p tons of potatoes, and he sells x_1 tons at the end of the first year. The yield from p tons is mp tons, and hence the amount remaining for the second year is $(mp - x_1)$ tons; but this amount must be used optimally in the remaining $(n-1)$ years, and so the farmer's maximum total income is

$$B_n(p) = \max_{x_1}\{f(x_1) + B_{n-1}(mp - x_1)\}, \tag{6.31}$$

where

$$0 \leqslant x_1 \leqslant mp.$$

The state variable after one year is

$$\xi_1 = mp - x_1 \geqslant 0, \tag{6.32}$$

and hence

$$\xi_1 \leqslant mp.$$

Repeating the above argument for the $(n-1)$-stage problem—starting the second year with ξ_1 tons and selling x_2 tons at the end of that year, etc.—we obtain

$$B_{n-1}(\xi_1) = \max_{x_2}\{f(x_2) + B_{n-2}(m\xi_1 - x_2)\}, \tag{6.33}$$

where

$$0 \leqslant x_2 \leqslant m\xi_1.$$

The state variable after two years is

$$\xi_2 = m\xi_1 - x_2 \geqslant 0,$$

and hence

$$\xi_2 \leqslant m\xi_1 \leqslant m^2 p.$$

Similarly, for the k-stage problem, we have

$$B_k(\xi_{n-k}) = \max_{x_{n-k+1}} \{f(x_{n-k+1}) + B_{k-1}(m\xi_{n-k} - x_{n-k+1})\}, \tag{6.34}$$

where

$$0 \leqslant x_{n-k+1} \leqslant m\xi_{n-k}. \tag{6.35}$$

Equation (6.34) gives the total income from the beginning of year $(n-k+1)$ to the end of year n. The state variable after $(n-k+1)$ years is

$$\xi_{n-k+1} = m\xi_{n-k} - x_{n-k+1} \geqslant 0, \tag{6.36}$$

and hence

$$\xi_{n-k+1} \leqslant m\xi_{n-k} \leqslant m^{n-k+1} p.$$

Setting $k = 2$ in equations (6.34) to (6.36), we obtain

$$B_2(\xi_{n-2}) = \max_{x_{n-1}} \{f(x_{n-1}) + B_1(m\xi_{n-2} - x_{n-1})\}, \tag{6.37}$$

where

$$0 \leqslant x_{n-1} \leqslant m\xi_{n-2};$$

and

$$\xi_{n-1} = m\xi_{n-2} - x_{n-1} \geqslant 0,$$

with

$$\xi_{n-1} \leqslant m^{n-1} p.$$

Finally, for the 1-stage problem, we have

$$B_1(\xi_{n-1}) = f(m\xi_{n-1}), \tag{6.38}$$

because the given policy for the final year is to sell all the produce. Note that, in equation (6.38),

$$0 \leqslant m\xi_{n-1} \leqslant m^n p.$$

The recurrence relations (6.31), (6.33), (6.34), (6.37) and (6.38) can, of course, be written down in reverse order.

The solution of the problem is now obtained sequentially. Given the selling function f, the function B_1 is known from equation (6.38). Using f and B_1 in equation (6.37), the minimizing value of x_{n-1} can be found, and hence the function B_2 is known. Similarly, by using equation (6.34) successively with $k = 3,\ldots,n$, the functions $B_3,\ldots,B_n$ can be found. The farmer's maximum possible total income is then given by $B_n(p)$. In carrying out the above procedure, the optimal values of the decision variables are obtained in the order $x_{n-1}^*,\ldots,x_1^*$.

Example 6.4

Find the farmer's maximum income and his optimal selling policy when $f(x) \equiv Cx^s$ $(0 < s < 1)$ and $n = 3$.

Solution

From equation (6.38), we find

$$B_1(\xi_2) = Cm^s\xi_2{}^s.$$

Then equation (6.37) becomes

$$B_2(\xi_1) = \max_{x_2}\{Cx_2{}^s + Cm^s(m\xi_1 - x_2)^s\},$$

from which, by differentiation, the maximizing value of x_2 is found to be

$$\bar{x}_2 = \frac{m^{\frac{2s-1}{s-1}}\xi_1}{1 + m^{\frac{s}{s-1}}}, \tag{6.39}$$

and hence

$$B_2(\xi_1) = Cm^{2s}\left(1 + m^{\frac{s}{s-1}}\right)^{1-s}\xi_1{}^s.$$

Equation (6.31) now becomes

$$B_3(p) = \max_{x_1}\{Cx_1{}^s + Cm^{2s}\left(1 + m^{\frac{s}{s-1}}\right)^{1-s}(mp - x_1)^s\}, \tag{6.40}$$

which gives, by differentiation,

$$x_1{}^* = \frac{m^{\frac{3s-1}{s-1}}p}{1 + m^{\frac{s}{s-1}} + m^{\frac{2s}{s-1}}}. \tag{6.41}$$

Substituting this value of x_1 in equation (6.40) gives the farmer's maximum income:

$$B_3(p) = Cm^{3s}\left(1 + m^{\frac{s}{s-1}} + m^{\frac{2s}{s-1}}\right)^{1-s}p^s.$$

The optimal selling policy is given by $x_1{}^*$ of equation (6.41), together with

$$x_2{}^* = \frac{m^{\frac{2s-1}{s-1}}\xi_1{}^*}{1 + m^{\frac{s}{s-1}}},$$

the latter coming from equation (6.39) with $\xi_1{}^*$ substituted for ξ_1. From equations (6.32) and (6.41), we find

$$\xi_1{}^* = \frac{1 + m^{\frac{s}{s-1}}\, mp}{1 + m^{\frac{s}{s-1}} + m^{\frac{2s}{s-1}}},$$

and hence

$$x_2{}^* = \frac{m^{\frac{3s-2}{s-1}}\, p}{1 + m^{\frac{s}{s-1}} + m^{\frac{2s}{s-1}}}.$$

6.6 SCHEDULING PROBLEMS

Many realistic industrial and commercial optimization problems take the form of scheduling problems, i.e. problems in which a sequence of operations has to be planned in such a way that a given objective function is maximized or minimized. Typical examples are:

(a) The servicing, repair and replacement of machinery.

(b) The most efficient use of several machines which are used sequentially to produce manufactured articles.

(c) The problem of stock control in warehouses.

(d) The 'caterer problem': what is the optimal policy for a caterer who wishes to minimize the cost of providing clean table napkins for n days if he knows the daily demand, the purchase price and the costs of a fast and a slow laundry service?

(e) The 'travelling salesman problem': what is the shortest route which passes once and only once through each of n given cities?

Detailed formulations of these problems may be found in many texts.[5,6,57,76] For most scheduling problems there are alternative methods of solution, not all of which use dynamic programming.

In the remainder of this section we shall consider the simplest case of problem (b), namely, the optimal scheduling of two machines A and B on which n articles have to be processed. Each article has to pass through machine A and then through machine B, only one article at a time can go through a machine, and the processing times for each article on each machine are known. The problem is to find the order in which the n articles should be fed to machine A so that the total processing time is minimized.

The corresponding problem with three machines is much more difficult to solve; the explicit dynamic-programming solution for the two-machine problem cannot be extended to the three-machine problem. The latter problem can, however, be solved by the technique of permutation programming, which was developed by Nicholson.[58] A much more complicated problem of this type has been solved by Nicholson and Pullen.[59]

Returning to the two-machine problem, let a_j, b_j be the times required to process the jth article on machines A and B, respectively. Define the return function

$$R_n(a_1,b_1,....,a_n,b_n;\tau)$$

as the time taken to process the n articles, starting from a state in which machine B is fully committed for the next τ units of time and an optimal scheduling policy is followed. In this definition, the commitment of machine B is quite arbitrary: it may be due to jobs not concerned with the present n articles, and could even include tea breaks! Note that the state variable τ in the definition of R_n implies a generalization of the original problem.

Suppose that the pth article is the first to be fed to machine A. Then, considering the processing of the pth article separately from the rest, we must have

$$R_n(a_1,b_1,....,a_n,b_n;\tau) = \min_{p=1,....,n} \{a_p + R_{n-1}(a_1,b_1,....,0,0,....,a_n,b_n;\tau_p\},$$

where the numbers 0,0 on the right replace a_p,b_p on the left, and

$$\tau_p = b_p + \max\{(\tau - a_p), 0\}. \tag{6.42}$$

To explain the form (6.42) for the state variable τ_p, suppose that the processing of the pth article has just been completed by machine A. At this time, machine B is fully committed for a further time τ_p, given by

$$\left.\begin{aligned} \tau_p &= b_p & \text{if} \quad a_p \geqslant \tau, \\ &= b_p + \tau - a_p & \text{if} \quad a_p < \tau. \end{aligned}\right\} \tag{6.43}$$

Equation (6.42) is merely a concise form of equations (6.43).

The general idea of the solution is that, given any permutation of the articles, we consider whether the value of the objective function can be improved by interchanging a pair of adjacent articles, say the pth and the qth. This process is repeated until an optimal solution is found.

Suppose that the pth article is fed to machine A, followed immediately by the qth article. Then

$$\begin{aligned} R_n(a_1,b_1,....,a_n,b_n;\tau) = a_p + a_q + R_{n-2}(a_1,b_1,....,0,0,....,0,0,.... \\ ....,a_n,b_n;\tau_{pq}), \end{aligned} \tag{6.44}$$

where τ_{pq} is obtained by applying equation (6.42) twice:

$$\begin{aligned} \tau_p &= b_p + \max\{(\tau - a_p), 0\}, \\ \tau_{pq} &= b_q + \max\{(\tau_p - a_q), 0\}. \end{aligned}$$

Substituting for τ_p from the first of these equations in the second, we find

$$\begin{aligned} \tau_{pq} &= b_q + \max\{b_p - a_q + \max(\tau - a_p, 0), 0\} \\ &= b_p + b_q - a_q + \max\{\max(\tau - a_p, 0), a_q - b_p\} \\ &= b_p + b_q - a_p - a_q + \max\{\max(\tau, a_p), a_p + a_q - b_p\} \\ &= b_p + b_q - a_p - a_q + \max\{\tau, a_p, a_p + a_q - b_p\}. \end{aligned} \tag{6.45}$$

Similarly,

$$\tau_{qp} = b_q + b_p - a_q - a_p + \max\{\tau, a_q, a_q + a_p - b_q\}. \tag{6.46}$$

By definition, the value of the return function R_{n-2} of equation (6.44) decreases as τ_{pq} decreases, and hence we must make τ_{pq} as small as possible. We therefore place the pth and qth articles in the order (q,p) rather than (p,q) if $\tau_{qp} < \tau_{pq}$, i.e. if

$$\max\{a_q, a_q + a_p - b_q\} < \max\{a_p, a_p + a_q - b_p\}, \tag{6.47}$$

using equations (6.45) and (6.46). In writing down the inequality (6.47) we have assumed, without loss of generality, that

$$\tau < \max\{a_p, a_q, a_p + a_q - b_p, a_q + a_p - b_q\}.$$

The inequality (6.47) may be written

$$\min\{a_p, b_q\} > \min\{a_q, b_p\},$$

which leads to the following simple algorithm for the optimal scheduling of the two machines A and B, given the $2n$ processing times a_j, b_j:

1. Find by inspection the minimum processing time $\min_j\{a_j, b_j\}$.

2. If the minimum in step 1 is an a value, process the article first; if a b value, last.

3. Repeat steps 1 and 2 with the reduced set of $(2n-2)$ processing times. Continue until all the articles have been placed in order.

4. Ties for the minimum processing time may be resolved by using the first applicable rule of the following three:

(a) If $a_p = a_q$, choose article p to be first if $p < q$.
(b) If $b_p = b_q$, choose article p to be last if $p < q$.
(c) If $a_p - b_q$, choose article p to be first.

Note that the existence of ties implies the existence of alternative optimal schedules.

Example 6.5

The processing times a_j, b_j for eight articles on two machines A and B are given in Table 6.5. The articles must pass through machine A and then through machine B. Each machine can process only one article at a time. Find an optimal schedule for the processing of the articles.

Table 6.5 Processing times for eight articles on machines A and B

j	1	2	3	4	5	6	7	8
a_j	12	2	4	19	19	16	19	16
b_j	2	10	10	13	5	7	2	20

Solution

Tables 6.6(i) to (vii) show the results of successive steps of the above algorithm. When an article has been placed in order, its number j is written in bold type.

Table 6.6 Solution of Example 6.5

(i)	j	2	1	3	4	5	6	7	8
	a_j	2	12	4	19	19	16	19	16
	b_j	10	2	10	13	5	7	2	20
(ii)	j	2	3	4	5	6	7	8	1
	a_j	2	4	19	19	16	19	16	12
	b_j	10	10	13	5	7	2	20	2
(iii)	j	2	3	4	5	6	8	7	1
	a_j	2	4	19	19	16	16	19	12
	b_j	10	10	13	5	7	20	2	2
(iv)	j	2	3	4	5	6	8	7	1
	a_j	2	4	19	19	16	16	19	12
	b_j	10	10	13	5	7	20	2	2
(v)	j	2	3	4	6	8	5	7	1
	a_j	2	4	19	16	16	19	19	12
	b_j	10	10	13	7	20	5	2	2
(vi)	j	2	3	4	8	6	5	7	1
	a_j	2	4	19	16	16	19	19	12
	b_j	10	10	13	20	7	5	2	2
(vii)	j	2	3	8	4	6	5	7	1
	a_j	2	4	16	19	16	19	19	12
	b_j	10	10	20	13	7	5	2	2

Table 6.6(vii) is optimal. The articles should be fed to machine A in the order 2, 3, 8, 4, 6, 5, 7, 1.

SUMMARY

Dynamic programming is an optimization technique based on Bellman's 'Principle of Optimality'. It may be used to solve problems that require a sequence of decisions to be taken, provided that when any decision is taken the remaining sub-problem has the same structure as the previous sub-problem. Many problems which do not at first sight fall into this class can be converted into dynamic programming form by imbedding them within a family of similar problems and by choosing suitable state and decision variables. Sections 6.2 to 6.6 illustrate this procedure in a variety of situations. Unlike most other optimization techniques, dynamic programming tends to become computationally simpler as more constraints are added. In particular, it is often very useful when the variables are restricted to being integers. On the other hand, the computation of solutions can be prohibitively lengthy when the number of state variables at each stage exceeds three or four.

EXERCISES

1. Maximize $z = 2x_1^2 + 3x_1 - x_2^2 + 4x_2 + 5x_3^2 - 3x_3$, subject to

$$3x_1^2 + 2x_2^2 + x_3^2 \leqslant 15,$$
$$x_1, x_2, x_3 \text{ non-negative integers.}$$

2. Solve by dynamic programming and also by direct tabulation:

$$\text{maximize} \quad z = (x_1^2 - x_1)(x_2^2 - 2x_2)(x_3^2 - 3x_3),$$

subject to

$$x_1^2 + 2x_2^2 + 3x_3^2 \leqslant 100,$$
$$x_1, x_2, x_3 \text{ non-negative integers.}$$

3. Show how to solve the allocation problem (Section 6.2) when the inequality constraint (6.2) is replaced by an equation. Illustrate the method by solving Example 6.1 in this case.

Find the paths of minimum length from $x = 0$ to $x = 4$ in the networks of Exercises 4 and 5.

4.

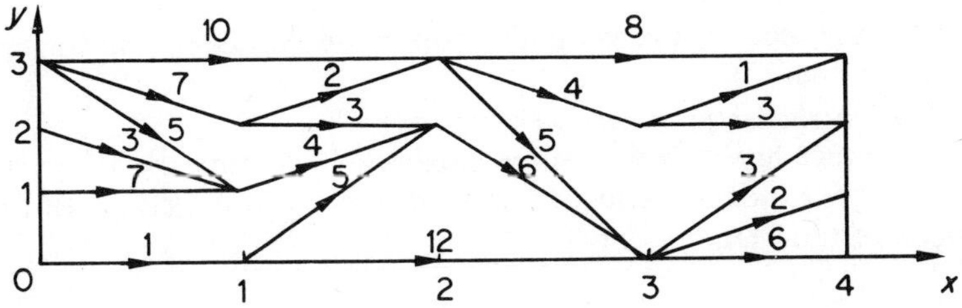

5.

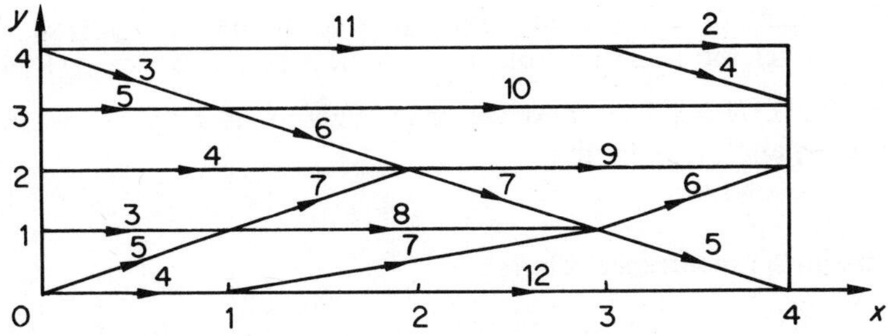

Find the paths of minimum length from node 1 to node 8 in the networks of Exercises 6 and 7.

6. 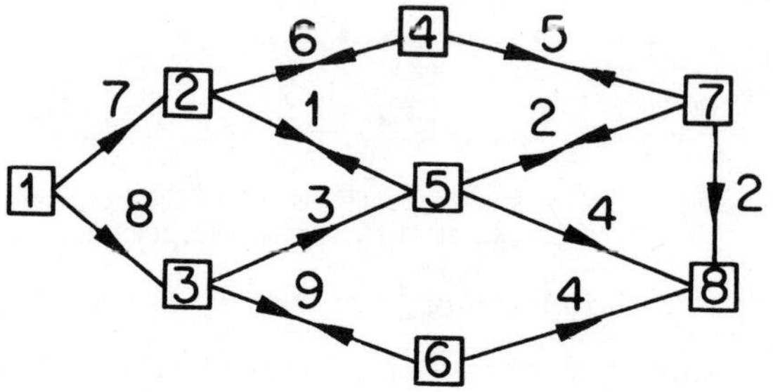

7. 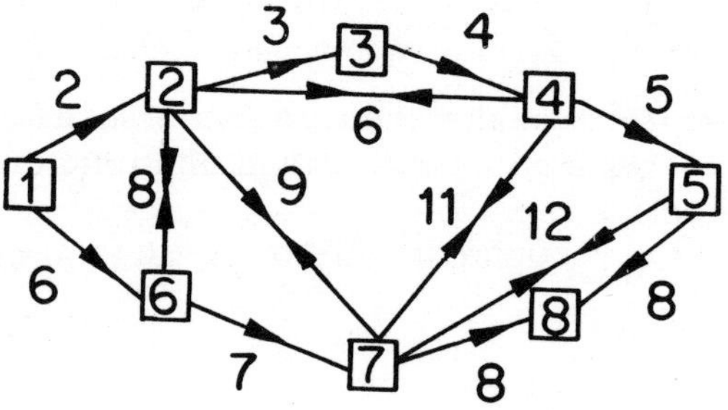

8. Find the second shortest paths from node 1 to node 8 in the networks of Exercises 6 and 7.

9. *The warehouse problem*

A warehouse has a fixed storage capacity C. At time $t = 0$ the stock is V; subsequently, goods are bought and sold in each of N periods of time. When i periods of time remain, let

$$\begin{aligned} c_i &= \text{cost price per unit,} \\ s_i &= \text{selling price per unit,} \\ x_i &= \text{quantity bought,} \\ y_i &= \text{quantity sold.} \end{aligned}$$

Goods bought in any period cannot be sold in the same period. The total profit P_N over the N periods of time is to be maximized. Let $f_N(V) = \max_{x_i, y_i} P_N$, where $V \geqslant 0$, $N = 1, 2, \ldots.$, and the maximization is taken over all possible values of x_i and y_i. Show that

$$f_1(V) = s_1 V,$$

and obtain the recurrence relation

$$f_N(V) = \max_{x_N, y_N} \{s_N y_N - c_N x_N + f_{N-1}(V + x_N - y_N)\}, \quad N \geqslant 2,$$

where

$$0 \leqslant y_N \leqslant V, \quad x_N \geqslant 0, \quad V + x_N - y_N \leqslant C.$$

10. In the farmer's problem (Section 6.5), assume that in any year there is a fixed probability λ that 10 per cent of the potatoes the farmer keeps for seed are destroyed by frost. Prove that the farmer's expected total income for the case of Example 6.4 becomes

$$CKm^s\left(1 + K^{\frac{1}{s-1}}\right)^{1-s} p^s,$$

where

$$K = m^{2s}\mu^s\left[\mu^s + (m\mu)^{\frac{s}{s-1}}\right]\left[1 + (m\mu)^{\frac{s}{s-1}}\right]^{-s},$$

and

$$\mu = 1 - 0{\cdot}1\lambda.$$

11. In the two-machine scheduling problem (Section 6.6), show that if the articles must pass through the machines in the order B–A, they should be fed to machine B in precisely the reverse order from that given by the A–B algorithm.

12. Find an optimal schedule for the processing of ten articles on two machines A and B, given the following processing times (the conditions of Example 6.5 apply).

j	1	2	3	4	5	6	7	8	9	10
a_j	43	75	31	25	18	75	6	25	6	25
b_j	27	8	14	58	2	8	10	42	22	92

References

1. J.C. Allwright, 'Optimal control synthesis using function decomposition techniques', *Proc. 4th IFAC Congress*, Warsaw, 1969.
2. Y. Bard, 'Comparison of gradient methods for the solution of nonlinear parameter estimation problems', *SIAM Journal on Numerical Analysis*, **7**, 157–186, 1970.
3. W.J. Baumol, *Economic Theory and Operations Analysis*, 3rd ed., Prentice-Hall, Englewood Cliffs, New Jersey, 1972.
4. E.M.L. Beale, 'Numerical methods', in *Nonlinear Programming* (Ed. J. Abadie), North-Holland, Amsterdam, 1967, Chap. VII.
5. R. Bellman, *Dynamic Programming*, Princeton University Press, Princeton, New Jersey, 1957.
6. R. Bellman and S.E. Dreyfus, *Applied Dynamic Programming*, Princeton University Press, Princeton, New Jersey, 1962.
7. M.C. Biggs, 'Minimization algorithms making use of non-quadratic properties of the objective function', *J. Inst. Maths. Applics.*, **8**, 315–327, 1971.
8. M.J. Box, 'A new method of constrained optimization and a comparison with other methods', *The Computer Journal*, **8**, 42–52, 1965.
9. M.J. Box, 'A comparison of several current optimization methods, and the use of transformations in constrained optimization', *The Computer Journal*, **9**, 67–77, 1966.
10. M.J. Box, D. Davies and W.H. Swann, *Non-linear Optimization Techniques*, I.C.I. Ltd., Monograph No. 5, Oliver and Boyd, 1969.
11. C.G. Broyden, 'Quasi-Newton methods and their application to function minimization', *Mathematics of Computation*, **21**, 368–381, 1967.
12. C.G. Broyden, 'A new method of solving nonlinear simultaneous equations', *The Computer Journal*, **12**, 95–99, 1969.
13. C.G. Broyden, 'The convergence of single-rank quasi-Newton methods', *Mathematics of Computation*, **24**, 365–382, 1970.
14. C.G. Broyden, 'The convergence of a class of double-rank minimization algorithms. 1. General considerations', *J. Inst. Maths. Applics.*, **6**, 76–90, 1970.
15. C.G. Broyden, 'The convergence of a class of double-rank minimization algorithms. 2. The new algorithm', *J. Inst. Maths. Applics.*, **6**, 222–231, 1970.
16. C.W. Carroll, 'The created response surface technique for optimising nonlinear restrained systems', *Operations Research*, **9**, 169–184, 1961.
17. A. Charnes and W.W. Cooper, *Management Models and Industrial Applications of Linear Programming*, Wiley, New York, 1961.
18. A. Charnes and C. Lemke, 'Minimization of nonlinear separable convex functionals', *Naval Research Logistics Quarterly*, **1**, 301–312, 1954.
19. R. Courant, *Differential and Integral Calculus*, Vol. II, Wiley-Interscience, New York, 1936.

20. R. Courant and D. Hilbert, *Methods of Mathematical Physics*, Vol. I, Interscience, New York, 1953.
21. G.B. Dantzig, 'Recent advances in linear programming', *Management Science*, **2,** 131–144, 1956.
22. G.B. Dantzig, 'On the status of multistage linear programming problems', *Management Science*, **6,** 53–72, 1959.
23. G.B. Dantzig, *Linear Programming and Extensions*, Princeton University Press, Princeton, New Jersey, 1963.
24. W.C. Davidon, 'Variable metric method for minimization', *A.E.C. R.&D. Report ANL*-5990(*Rev.*), Argonne National Laboratory, 1959.
25. D. Davies, 'The use of Davidon's method in nonlinear programming', *I.C.I. Ltd., Management Services Report MSDH*/68/110, 1968.
26. L.C.W. Dixon, 'Variable metric algorithms: necessary and sufficient conditions for identical behaviour of nonquadratic functions', *J. of Optimization Theory and Applications*, **10,** 34–40, 1972.
27. W.L. Ferrar, *Algebra* Oxford University Press, 1941.
28. A.V. Fiacco and G.P. McCormick, 'The sequential unconstrained minimization technique for nonlinear programming, a primal-dual method', *Management Science*, **10,** 360–366, 1964.
29. A.V. Fiacco and G.P. McCormick, 'Computational algorithm for the sequential unconstrained minimization technique for nonlinear programming', *Management Science*, **10,** 601–617, 1964.
30. A.V. Fiacco and G.P. McCormick, 'Extensions of SUMT for nonlinear programming: equality constraints and extrapolation', *Management Science*, **12,** 816–828, 1966.
31. A.V. Fiacco and G.P. McCormick, *Nonlinear Programming—Sequential Unconstrained Minimization Technique*, Wiley, New York, 1968.
32. R. Fletcher, 'Function minimization without evaluating derivatives—a review', *The Computer Journal*, **8,** 33–41, 1965.
33. R. Fletcher, 'Programming under linear equality and inequality constraints', *I.C.I. Ltd., Management Services Report MSDH*/68/19, 1968.
34. R. Fletcher, 'A new approach to variable metric algorithms', *The Computer Journal*, **13,** 317–322, 1970.
35. R. Fletcher, 'A general quadratic programming algorithm', *J. Inst. Maths. Applics.*, **7,** 76–91, 1971.
36. R. Fletcher and M.J.D. Powell, 'A rapidly convergent descent method for minization', *The Computer Journal*, **6,** 163–168, 1963.
37. R. Fletcher and C.M. Reeves, 'Function minimization by conjugate gradients', *The Computer Journal*, **7,** 149–154, 1964.
38. D. Gale, *The Theory of Linear Economic Models*, McGraw-Hill, New York, 1960.
39. P.E. Gill and W. Murray, 'Quasi-Newton methods for unconstrained optimization', *J. Inst. Maths. Applics.*, **9,** 91–108, 1972.
40. D. Goldfarb, 'Extension of Davidon's variable metric method to maximization under linear inequality and equality constraints', *SIAM Journal on Applied Mathematics*, **17,** 739–764, 1969.
41. D. Goldfarb and L. Lapidus, 'Conjugate gradient method for nonlinear programming problems with linear constraints', *Indust. and Eng. Chem. Fundamentals*, **7,** 142–151, 1968.
42. R.E. Griffith and R.A. Stewart, 'A nonlinear programming technique for the optimization of continuous processing systems', *Management Science*, **7,** 379–392, 1961.
43. G. Hadley, *Linear Programming*, Addison-Wesley, Reading, Mass., 1962.
44. G. Hadley, *Nonlinear and Dynamic Programming*, Addison-Wesley, Reading, Mass., 1964.
45. H. Hancock, *Theory of Maxima and Minima*, Dover, New York, 1960.
46. R. Hooke and T.A. Jeeves, '"Direct search" solution of numerical and statistical

problems', *J. of the Assn. for Computing Machinery*, **8,** 212–229, 1961.

47. H.Y. Huang, 'Unified approach to quadratically convergent algorithms for function minimization', *J. of Optimization Theory and Applications*, **5,** 405–423, 1970.
48. H.Y. Huang and A.V. Levy, 'Numerical experiments on quadratically convergent algorithms for function minimization', *J. of Optimization Theory and Applications*, **6,** 269–282, 1970.
49. H.J. Kelley, 'Methods of gradients', in *Optimization Techniques* (Ed. G. Leitmann), Academic Press, New York, 1962, Chap. 6.
50. J. Kiefer, 'Sequential minimax search for a maximum', *Proc. American Mathematical Society*, **4,** 502–506, 1953.
51. H.W. Kuhn and A.W. Tucker, 'Nonlinear programming', in *Proceedings of the Second Berkeley Symposium on Mathematical Statistics and Probability* (Ed. J. Neyman), University of California Press, Berkeley, 1951, pp. 481–492.
52. L.S. Lasdon, S.K. Mitter and A.D. Waren, 'The conjugate gradient method for optimal control problems', *I.E.E.E. Trans. Auto. Control*, Vol. AC-12, No. 2, 132–138, 1967.
53. K. Levenberg, 'A method for the solution of certain nonlinear problems in least squares', *Quart. Appl. Maths.*, **2,** 164–168, 1944.
54. D.W. Marquardt, 'An algorithm for least-squares estimation of nonlinear parameters', *J. Soc. Indust. Appl. Maths.*, **11,** 431–441, 1963.
55. W. Murray (Ed.), *Numerical Methods for Unconstrained Optimization*, Academic Press, London, 1972.
56. J.A. Nelder and R. Mead, 'A simplex method for function minimization', *The Computer Journal*, **7,** 308–313, 1965.
57. G.L. Nemhauser, *Introduction to Dynamic Programming*, Wiley, New York, 1966.
58. T.A.J. Nicholson, 'Optimizing problems of arrangements: permutation programming', *A.E.R.E. Theoretical Physics Division, Report No. T.P.* 309, 1967.
59. T.A.J. Nicholson and R.D. Pullen, 'A permutation procedure for job-shop scheduling', *A.E.R.E. Theoretical Physics Division, Report No. T.P.* 297, 1967.
60. J.D. Pearson, 'Variable metric methods of minimization', *The Computer Journal*, **12,** 171–178, 1969.
61. M.J.D. Powell, 'An iterative method for finding stationary values of a function of several variables', *The Computer Journal*, **5,** 147–151, 1962.
62. M.J.D. Powell, 'An efficient method of finding the minimum of a function of several variables without calculating derivatives', *The Computer Journal*, **7,** 155–162, 1964.
63. M.J.D. Powell, 'A method for minimizing a sum of squares of nonlinear functions without calculating derivatives', *The Computer Journal*, **7,** 303–307, 1965.
64. M.J.D. Powell, 'On the convergence of the variable metric algorithm', *J. Inst. Maths. Applics.*, **7,** 21–36, 1971.
65. M.J.D. Powell, 'Quadratic termination properties of minimization algorithms. I Statement and discussion of results', *J. Inst. Maths. Applics.*, **10,** 333–342, 1972.
66. M.J.D. Powell, 'Quadratic termination properties of minimization algorithms. II Proofs of theorems', *J. Inst. Maths. Applics.*, **10,** 343–357, 1972.
67. V. Riley and S.I. Gass, *Linear Programming and Associated Techniques: a Comprehensive Bibliography on Linear, Nonlinear and Dynamic Programming*, Johns Hopkins University Press, Baltimore, 1958.
68. S.M. Roberts and H.I. Lyvers, 'The gradient method in process control', *Indust. and Eng. Chem.*, **53,** 877–882, 1961.
69. J.B. Rosen, 'The gradient projection method for nonlinear programming. Part I. Linear constraints', *J. Soc. Indust. Appl. Maths.*, **8,** 181–217, 1960.
70. J.B. Rosen, 'The gradient projection method for nonlinear programming. Part II. Nonlinear constraints', *J. Soc. Indust. Appl. Maths.*, **9,** 514–532, 1961.
71. H.H. Rosenbrock, 'An automatic method for finding the greatest or least value of a function', *The Computer Journal*, **3,** 175–184, 1960.

72. C.S. Smith, 'The automatic computation of maximum likelihood estimates', *N.C.B. Scientific Dept., Report No. SC846/MR/40*, 1962.
73. W. Spendley, G.R. Hext and F.R. Himsworth, 'Sequential applications of simplex designs in optimization and evolutionary operation', *Technometrics*, **4**, 441–461, 1962.
74. W.H. Swann, 'Report on the development of a new direct search method of optimization', *I.C.I. Ltd., Central Instrument Laboratory Research Note* 64/3, 1964.
75. G.R. Walsh, *An Introduction to Linear Programming*, Holt, Rinehart and Winston, London, 1971.
76. D.J. White, *Dynamic Programming*, Oliver and Boyd, Edinburgh, 1969.
77. E.T. Whittaker and G.N. Watson, *Modern Analysis*, 4th ed., Cambridge University Press, 1950.
78. P. Wolfe, 'The simplex method for quadratic programming', *Econometrica*, **27**, 382–398, 1959.
79. P. Wolfe, 'Methods of nonlinear programming', in *Recent Advances in Mathematical Programming* (Eds. R.L. Graves and P. Wolfe), McGraw-Hill, New York, 1963, pp. 67–86.
80. L.C. Young, *Lectures on the Calculus of Variations and Optimal Control Theory*, Saunders, Philadelphia, 1969.

Suggestions for Further Reading

M. Athans and P.L. Falb, *Optimal Control: An Introduction to the Theory and Its Applications*, McGraw-Hill, New York, 1966.

A.V. Balakrishnan and L.W. Neustadt (Eds.), *Techniques of Optimization*, Academic Press, New York, 1972.

E.M.L. Beale (Ed.), *Applications of Mathematical Programming Techniques*, The English Universities Press, London, 1970.

A.E. Bryson and Y.C. Ho, *Applied Optimal Control: Optimization, Estimation and Control*, Blaisdell, Waltham, Mass., 1969.

R. Fletcher (Ed.), *Optimization*, Academic Press, New York, 1969.

L.S. Lasdon, *Optimization Theory for Large Systems*, Macmillan, New York, 1970.

D.A. Pierre, *Optimization Theory with Applications*, Wiley, New York, 1969.

L.S. Pontryagin, V.G. Boltyanskii, R.V. Gamkrelidze and E.F. Mishchenko, *The Mathematical Theory of Optimal Processes*, Interscience, New York, 1962.

P. Whittle, *Optimization under Constraints*, Wiley-Interscience, London, 1971.

Answers to Exercises

CHAPTER 1

1. (a) $\mathbf{x}^* = [\pm\sqrt{2}, \pm\sqrt{2}]$ (2 points); $z^* = 4$ (min).
   (b) $\mathbf{x}^* = [\pm 1, \pm 1]$ (4 points); $z^* = +1$ (max) or -1 (min).
2. $\mathbf{x}^* = [4/3, 1/3, 0]$; $z^* = 17/9$ (max);
   $\mathbf{x}^* = [0, 1, 0]$; $z^* = 1$ (min).
3. $a^2/2$.
5. $z^* = |\mathbf{a}|$.
6. $z^* = n^{-n/2}$. Set $a_j = nAx_j^2$, where the a_j are the given positive numbers and A is their arithmetic mean.
7. $A^*, B^*, C^* = 2k_1\pi, (2k_2 + h)\pi, -(2k_1 + 2k_2 + h)\pi$; $z^* = 1$ (max);
   $A^*, B^*, C^* = (2k_1 \pm \frac{1}{3})\pi, (2k_2 \pm \frac{1}{3})\pi, -(2k_1 + 2k_2 \pm \frac{2}{3})\pi$; $z^* = -\frac{1}{8}$(min);
   $A^*, B^*, C^* = (2k_1 \pm \frac{2}{3})\pi, (2k_2 \pm \frac{2}{3})\pi, -(2k_1 + 2k_2 \pm \frac{4}{3})\pi$; $z^* = -\frac{1}{8}$(min);
   where k_1 and k_2 are any integers, and $h = 0$ or 1.
10. $r_1 - \sqrt{2}a$, if $\theta_1 = 0$ or π *and* $r_1 \geqslant \sqrt{2}a$;
    $r_1 \left|\frac{\sin(\theta - \theta_1)}{\sin 2\theta}\right|$ otherwise, where θ is given by

    $$r_1 \sin(3\theta - \theta_1) = a\sqrt{2\cos 2\theta}\,\sin 2\theta,$$

    in which θ and θ_1 lie in the same quadrant, and $\cos 2\theta \geqslant 0$.
12. $\mathbf{x}^* = [4/5, 0]$; $z^* = -24/25$.
13. $\mathbf{x}^* = [1{\cdot}258, 3{\cdot}135, 1{\cdot}895]$; $z^* = 7{\cdot}47$.
14. Let α $(= 1{\cdot}197)$ be the real root of $2x^3 - x^2 - 2 = 0$.
    $k < 0$: z unbounded;
    $0 < k \leqslant \alpha$: $x^* = \alpha$; $z^* = -1{\cdot}907$;
    $k \geqslant \alpha$; $x^* = k$.
15. (a) Convex.
    (b) Convex.
    (c) Neither convex nor concave in general.
    (d) Concave.
    (e) Convex.
19. There is no function satisfying conditions (d).
21. (a) $\mathbf{A}$ is negative semidefinite.
    (b) $\mathbf{A}$ is positive definite.
    (c) $\mathbf{A}$ is negative definite.
22. $(\mathbf{x}_2 - \mathbf{x}_1)'\mathbf{A}(\mathbf{x}_2 - \mathbf{x}_1) > 0$ for all $\mathbf{x}_1, \mathbf{x}_2 (\neq \mathbf{x}_1) \in X$.

CHAPTER 2

1. $\mathbf{x}^* = [6,0]$; $z^* = 36$. The sufficient conditions of Theorem 2.4 are not satisfied.
2. Let $\mathbf{X} = \mathbf{x} - \mathbf{a}$; then the problem in $\mathbf{X}$ can be written in the form (2.1).
3. No, because the vectors $\nabla g_1(\mathbf{x}^*)$ and $\mathbf{e}_2$ are linearly dependent (see page 46).
4. K–T I: $\mathbf{A}'\boldsymbol{\lambda}^* \geqslant \mathbf{c}$; K–T II: $(\mathbf{c} - \mathbf{A}'\boldsymbol{\lambda}^*)'\mathbf{x}^* = 0$;
   K–T III: $\mathbf{A}\mathbf{x}^* \leqslant \mathbf{b}$; K–T IV: $(\mathbf{b} - \mathbf{A}\mathbf{x}^*)'\boldsymbol{\lambda}^* = 0$.
   The dual of the given linear programming problem is:

   $$\text{minimize } w = \mathbf{b}'\boldsymbol{\lambda}, \text{ subject to } \mathbf{A}'\boldsymbol{\lambda} \geqslant \mathbf{c},\ \boldsymbol{\lambda} \geqslant \mathbf{0}.$$

   Thus K–T I and III give the dual and primal constraints, respectively; and K–T II and IV together show that $z^* = w^*$.
6. No. The optimal points are $[a,0]$ and $[0,a]$; the constraint function is not differentiable at either of these points.
7. (i) $\mathbf{x}^* = [0,1]$; none of the conditions satisfied.
   (ii) as for (i).
   (iii) $\mathbf{x}^* = [0,1]$; (a), (c), (d) satisfied, (b) not satisfied.
   $\mathbf{x}^* = [0,y]$, $0 \leqslant y < 1$; all four conditions satisfied.
8. The inequality (2.49) is restricted to the case where $\mathbf{dx} \geqslant \mathbf{0}$; equation (2.51) becomes a $\geqslant$ inequality; and the inequality (2.52) no longer becomes a strict equality for components $j \in J_p$.
9. $\mathbf{x}^* = [4/11, 3/11]$; $z^* = 5/11$.
10. $\mathbf{x}^* = [9/2, 7/2]$; $z^* = 53/4$.
11. $\mathbf{x}^* = [71/74, 83/74]$; $z^* = 897/148 = 6{\cdot}196$.
12. $\mathbf{x}^* = [12,0]$; $z^* = 120$. Not solvable by Wolfe's algorithm.
13. $\mathbf{x}^* = [7/10, 11/10]$; $z^* = 47/20$.
14. $(\mathbf{x}^*, \boldsymbol{\lambda}^*) = [7/10, 11/10, 0, 0]$; $w^* = 47/20$.
    $\boldsymbol{\lambda}^* = \mathbf{0}$ because both primal constraints are passive.
17. $\mathbf{x}_2 = [3/2, 3, 3]$.

CHAPTER 3

1. $[x^*, y^*] = [3/2, 2]$; $f(x^*, y^*) = 8e^{-7/2} = 0{\cdot}2416$.
3. $\mathbf{b}_1 = [0,0]$: $\mathbf{x}^* = [-1,0]$; $f(\mathbf{x}^*) = -5$ (exact solution).
   $\mathbf{b}_1 = [4,2]$: $\mathbf{x}^* = [2,2]$; $f(\mathbf{x}^*) = -4$. For greater accuracy, the final step length should be reduced in this case.
4. $\mathbf{x}^* = [0{\cdot}8, 0{\cdot}2]$; $f(\mathbf{x}^*) = -0{\cdot}08$.
   Exact solution: $\mathbf{x}^* = [5/6, 1/6]$; $f(\mathbf{x}^*) = -1/12$.
5. $\mathbf{x}^* = [5/8, 1]$; $f(\mathbf{x}^*) = -307/64 = -4{\cdot}797$.
   Exact solution: $\mathbf{x}^* = [4/17, 13/17]$; $f(\mathbf{x}^*) = -83/17 = -4{\cdot}882$.
7. Exact solution: $\mathbf{x}^* = [2,1]$; $f(\mathbf{x}^*) = -15$.
9. $x^* = 0{\cdot}860$, $f(x^*) = 0{\cdot}561$.
10. $x^* = 0{\cdot}06389$ $[L_{11} = 144, X^* = 101]$.
    Exact solution: $x^* = (7 - 3\sqrt{5})/4 = 0{\cdot}07295$.
11. $x^* = 0{\cdot}36$; $f(x^*) = 1{\cdot}174129$.
12. $X^* = 1500$; $\phi(X^*) = -1{\cdot}68735 \times 10^{12}$.
    $x^* = 15{\cdot}00$; $f(x^*) = -1{\cdot}68735 \times 10^4$.
14. $\mathbf{x}^* = [5{\cdot}898, 28{\cdot}941]$; $f(\mathbf{x}^*) = -977{\cdot}2$.
15. Distance along λ-axis from current point to minimizing point—if f is a quadratic function and if f_e is exact.
16. Iteration 4 is identical with iteration 2; the method therefore goes into an endless cycle. The cycle can be broken, however, by discarding the point $\alpha = 0$ and replacing it by the point $\alpha = -2{\cdot}5$ after iteration 4. This procedure is contrary to rule 5; but

the calculations then proceed normally.
17. Exact solution: $\mathbf{x}^* = [13/8, -3/2, -7/8]$.

CHAPTER 4

1. $\mathbf{x}^* = [-1{\cdot}380, -5{\cdot}261]$; $f(\mathbf{x}^*) = -20{\cdot}41$.
3. $\mathbf{x}^* = [2,3]$; $f(\mathbf{x}^*) = 1$.
5. $\mathbf{x}^* = [-1/4, -1/4]$; $f(\mathbf{x}^*) = -1/8$.
6. $\mathbf{x}^* = [1{\cdot}76923, 2{\cdot}26923, 2{\cdot}26923]$.
7. *Hint*: the given matrix can be reduced to diagonal form by a similarity transformation; and a diagonal matrix has an obvious square root.
9. $\mathbf{x}^* = [10/19, 13/19, 21/19] = [0{\cdot}5263, 0{\cdot}6842, 1{\cdot}1053]$; $f(\mathbf{x}^*) = 1$.
11. The statement is false in general: consider the case $\mathbf{G} = \mathbf{I}$, for example.
13. $x_1^* = \frac{1}{2}\sqrt{\frac{241}{30}} = 1{\cdot}4172$, $x_2^* = \frac{1}{10}$; $f(\mathbf{x}^*) = \frac{361}{480} - \frac{1}{2}\sqrt{\frac{241}{30}} = -0{\cdot}6651$.
14. $\mathbf{x}^* = [-2, -2]$; $f(\mathbf{x}^*) = 0$.
15. $\mathbf{x}^* = [-3, -1]$; $f(\mathbf{x}^*) = 0$.
16. $\mathbf{x}^* = [7/23, 4/23] = [0{\cdot}3043, 0{\cdot}1739]$; $f(\mathbf{x}^*) = 14/23 = 0{\cdot}6087$.
17. $T_0 = 288{\cdot}0°$ C, $\alpha = 0{\cdot}001981°$ C/ft.

CHAPTER 5

3. The negative gradient direction.
4. $[-2, -1]$.
5. $\mathbf{x}^* = [0,0,5]$; $z^* = -15$.
6. $[4,0,-1]$, $[675,5,-157]$.
7. Second approximation gives $[1,3]$; true value is $[1, 2{\cdot}9948]$.
8. True $x^* = 2{\cdot}02876$.
   (a) $x^* = 3{\cdot}46831$, error $= 1{\cdot}43955$.
   (b) $x^* = 2{\cdot}02687$, error $= -0{\cdot}00189$.
9. $\mathbf{x}^* = [0{\cdot}7590, 0{\cdot}2880]$; $f(\mathbf{x}^*) = -2{\cdot}148$.
10. $\mathbf{x}^* = \left[\sqrt{\frac{23}{10}}, \sqrt{23}, \frac{1}{2}\sqrt{23}\right] = [1{\cdot}517, 4{\cdot}796, 2{\cdot}398]$; $f(\mathbf{x}^*) = 17{\cdot}44$.
11.. $\mathbf{x}^* = [1{\cdot}9108, -0{\cdot}8216, -1{\cdot}9108]$ or $[-0{\cdot}8216, 1{\cdot}9108, -1{\cdot}9108]$; $f(\mathbf{x}^*) = 7{\cdot}977$.
12. All three constraints are active. The only solutions are $\mathbf{x} = [0{\cdot}5170, -1{\cdot}4684, -3{\cdot}9514]$ and $\mathbf{x} = [-1{\cdot}4684, 0{\cdot}5170, -3{\cdot}9514]$; $f(\mathbf{x}) = 18{\cdot}04$.
13. $\mathbf{x}^* = [1{\cdot}7818, 4{\cdot}4898]$; $f(\mathbf{x}^*) = 30{\cdot}24$.
14. $\mathbf{x}^* = [2\sqrt{10}/5, \sqrt{10}/5] = [1{\cdot}2649, 0{\cdot}6325]$; $f(\mathbf{x}^*) = 2$.
16. $\mathbf{x}^* = [0{\cdot}2, 0{\cdot}4]$; $f(\mathbf{x}^*) = 1{\cdot}48$.

CHAPTER 6

1. $\mathbf{x}^* = [1,2,2]$; $z^* = 23$.
2. $\mathbf{x}^* = [4,4,4]$; $z^* = 384$.
3. $\mathbf{x}^* = [3,2,0]$; $z^* = 13$.
4. Two minimal paths of length 14:
   $(0,0) \to (1,0) \to (2,2) \to (3,0) \to (4,1)$;
   $(0,3) \to (1,2) \to (2,3) \to (3,2) \to (4,4)$.
5. Three minimal paths of length 13:
   $(0,4) \to (3,4) \to (4,4)$;
   $(0,4) \to (1,3) \to (4,3)$;
   $(0,2) \to (2,2) \to (4,2)$.

6. Two minimal paths of length 12:
   $1 \to 2 \to 5 \to 7 \to 8$;
   $1 \to 2 \to 5 \to 8$.
7. One minimal path of length 19:
   $1 \to 2 \to 7 \to 8$.
8. Ex. 6: $1 \to 2 \to 5 \to 2 \to 5 \to 8$, length 14;
   Ex. 7: $1 \to 6 \to 7 \to 8$, length 21.
12. The articles should be fed to machine A in the order:
    7, 9, 4, 8, 10, 1, 3, 6, 2, 5.

Index

See also References (pp. 188–191) *and* Suggestions for Further Reading (p. 192)